T0313345

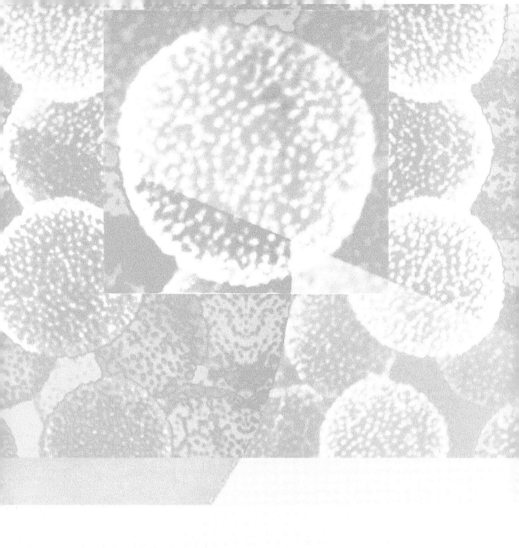

MOBILE MICROSPIES

MOBILE MICROSPIES

Particles for Sensing and Communication

Michael Köhler

PAN STANFORD PUBLISHING

Published by

Pan Stanford Publishing Pte. Ltd.
Penthouse Level, Suntec Tower 3
8 Temasek Boulevard
Singapore 038988

Email: editorial@panstanford.com
Web: www.panstanford.com

British Library Cataloguing-in-Publication Data
A catalogue record for this book is available from the British Library.

Mobile Microspies: Particles for Sensing and Communication

ISBN 978-981-4800-14-3 (Hardcover)
ISBN 978-0-429-44856-0 (eBook)

Contents

Preface

Freedom in combinations of a few types of atoms in molecular architectures supplies an unlimited variability in the organic chemical microcosmos. It is also the fundament for forming biomolecules that are able to construct and manage biological cells. These and other atoms are able to form solids, microparticles, and nanoparticles, too, which possess the ability to generate a huge variability in structures and functions by structural combination at the mesoscale between the level of molecules and the macroscopic world. There is not only "plenty of room" at the molecular scale, but there also is an unimaginably huge space for constructing objects between the nanometer and the millimeter scale.

During the past two decades, it was recognized by many researchers that a treasure of possibilities is offered by the world of micro- and nanoparticles. One of the most fascinating aspects is that their properties can be influenced not only by their chemical composition but also by their size, shape, and environment. New synthesis strategies, new laboratory techniques (among them the microfluidic method), and new application requirements led to a large spectrum of new synthetic particle types. Many of them are of interest for new applications because they are usable for local conversion of chemical information into physically readable signals. These particles can be mobile and can be implemented in different liquid and technical systems, such as in cells, tissues, or the environment, too. In communication with suitable readout systems, they can report about local chemical conditions and processes. That's why they can be regarded as "mobile microspies" (in German "mobile mikro-Spione").

After giving a talk at a meeting in San Diego in December 2016, the publisher asked me to write a book about *mobile sensor particles*. The topic is fascinating to me for about 25 years. In this time, some colleagues and I had worked mainly on using micro- and millifluidics for improving lab methods for the synthesis and development of new types of micro- and nanoparticles. For a fruitful collaboration on microfluidic synthesis and application of nanoparticles, in particular, I would like to thank Christophe Serra, Chenqi Chang, Wolfgang Fritzsche, Andrea Csaki, Jörg Reichert, Jörg Wagner, Andrea Knauer, Shuning Li, Steffen Schneider, Aniket Thete, Nikunjkumar Visaveliya, Stefan Nagl, Anette Funfak, Jialan Cao, Lars Hafermann, and Xiang Li. I also have to thank them for their cooperation, interesting and stimulating discussions, and essential support. The common work mainly motivated me to deal with the large world of small particles.

Michael Köhler

2018

Introduction

Particle-based sensing techniques have attracted quickly increasing interest during the past two decades. This development is driven by a broad spectrum of new technologies for the preparation and measurement of micro- and nanoparticles and by fascination of the possibilities for designing and functionalization of all specificities in the structure shapes and behavior of these tiny objects, on the one hand. On the other hand, there evolves a fast-growing need of new sensing and communication paths for medicine, biotechnology, and analytical science, as well as for new and efficient information transfer and storage systems, in general. No longer are particles regarded only as special types of materials, but it is better understood that they are bridging the gap between material and system, between structure and function. Thus, micro- and nanoparticles can act as transducers and can used for local signal conversion and for readout of information from any environment in which they are embedded.

The book introduces the concepts of bead-based sensing and "mobile spies" at the micro- and nanoscale. It gives an overview of the role of particles in contact signal conversion and noncontact signal transfer and of the interaction of these two processes. Therefore, a large spectrum of methods for coupling molecular recognition with primary signal transduction and optical signal transfer is presented. Besides, fluorescence and fluorescence resonance energy transfer (FRET) beads, microparticles using phosphorescence or chemoluminescence, photoacoustic transduction, pohotochemical switching, and bead-based plasmonic and Raman sensing, are discussed. But it becomes clear that optical signal transduction with visible light is not the only way for realizing microsensor particles. Ultraviolet (UV) and infrared (IR) waves, electrochemical functions, and magnetic functions can also be involved in realizing miniaturized and mobile transducers.

Synthesis and surface functionalization are key issues in the entire sensor particle development. The methods and procedures are dependent on particle materials; their functions, transport, and storage requirements; and the required particle sizes and homogeneity requirements. It is shown that techniques for the production of composite particles, hierarchically composed particles, and particles from microfluidic syntheses are of particular interest from the point of view of the need of high-quality sensing micro-objects.

The following chapter is devoted to the application of microscopic "spy"-like particles for information read out from different systems. These applications reach from technical microenvironments, biotechnology, and lab applications in biomedicine to environmental tasks. Meanwhile, an impressive part of investigations shows the applicability of a particle-based readout of different local concentrations of chemical species and biomolecules from organs, tissues, and single cells, too.

The use of particle-like objects for spreading information and communication is a not completely new invention in recent technology. The principle had already been developed by living nature many million years ago. Therefore, the analogies between particle-based information transfer in organisms and natural environments, on the one hand, and particle-based sensing and communication by technical systems, on the other hand, are discussed. It has to be recognized that a certain convergence of these technical developments in the direction of the natural principle is evident. This convergence is due to the efficiency of particle-based transduction, transport, and information storage principles. And it is also required for reaching an eco-compatible and sustainable application of these materials and technologies. Environmental requirements and the question of how we can interconnect better natural and technical information systems' demand for a new concept of particle-based technologies drive the search for new particle designs and particle-related technologies. The state of the art in the development of particles as "mobile spies," for communication and information management, allows us to speculate about future particle-based components and systems in a completely sustainable world economy.

In the last chapter of this book, some aspects of possible future developments are discussed.

Chapter 1

Challenge of Sensing

Advanced technical systems are not imaginable without sensors. Their ability to receive signals from the environment, to extract relevant data, and to monitor the changing of environmental conditions and the status of the technical system itself are essential for technical systems. Sensing is not only important for industrial automata, for cars and airplanes and nearly the whole spectrum of recent household devices, but is a fundamental feature of life. All living systems have to respond to their environment to get special orientation, to pick up nutrients, to look for partners and competitors, and to recognize any danger around.

It is typical for biological systems that different sensing functions are related to specially evolved recognition elements. Such elements are to find on the lower nanometer range on the level of single sensing molecules, on the micrometer scale in the form of recognition cells and cell organelles, and on the macroscopic scale in the form of sensing organs like ears and eyes. It is interesting to see that the tasks of specific sensing in nature are always connected with the formation and use of spatially separated units. This specialization follows the general principle of compartmentalizing in nature. It allows living beings to optimize the generation and application of all required information uptake and to combine lower, specific high-rate sensing of basic signal classes with highly specific sensing of selected signals with particular importance for their system. Multicellular organisms

Mobile Microspies: Particles for Sensing and Communication
Michael Köhler
Copyright © 2019 Pan Stanford Publishing Pte. Ltd.
ISBN 978-981-4800-14-3 (Hardcover), 978-0-429-44856-0 (eBook)
www.panstanford.com

have developed a lot of different sensing elements at all size levels. They enable the organisms to receive a lot of information by very different sensing channels. The ability to survive in a complex and permanently changing environment is strongly dependent on this receiving, processing, and evaluation of incoming information.

Older technical systems had worked without sensors or with a very low number of selected sensing elements. Often, the sensing ability of the operator, the system-supervising human being, is crucial for the function of the technical system. Classical car driving is not imaginable without the eyes—and to certain extent without the ears and the acceleration sensing—of the driver. Meanwhile, the sensing functions are more and more complemented or substituted by technical sensors for imaging, for temperature and pressure monitoring, for measuring inertial forces, and for controlling several other system parameters.

Despite the increasing installation of sensors, a miss ratio between the ability of computational power and sensing ability exists in most recent technical systems: millions or billions of transistors in the computer processor, on the one hand, can process data coming from some dozens or some hundreds of sensors only, on the other hand. It is expected that this ratio will shift in the next generations of technical devices. Future intelligent technical systems will have to know much more about their environment and their interactions on different levels. Thus, new strategies have to be developed to enable the systems to operate an increasing number of sensing channels and for equipping the systems with the required sensors. Probably, each future system will be confronted with the "big-data problem," and will have to solve the requirements of uptaking and evaluating a lot of different incoming signals. Therefore, new concepts of sensing have to be developed.

Chapter 2

Technical Concept: Particle-Based Sensing

2.1   Requirements

Many classical devices and systems are equipped with sensors. These sensors have their specific function and a fixed place inside the technical system. The technical environment of each sensor is stable, and the input and output channels as well as the power supply are well defined. Such types of sensors are very useful for measurement tasks, which can be defined clearly by the kind and strength of expected input signals, by the required position of the sensors inside the system, and the boundary conditions of operation. But the design and application of such sensors demand an accurate predefinition of sensor function and its technical embedding. The construction of the sensor has to reconsider the expected incoming signals and the required output data and placing of the sensor. These decisions demand a restrictive selection of expected input signals in the phase of designing the technical system, already.

Each technical and each natural system produce permanently a high number of signals that are released into the environment. These signals are very different in their physical nature and in their frequency, strength, and direction of release. And each technical as well as each living system is dependent on the recognition of signals

Mobile Microspies: Particles for Sensing and Communication
Michael Köhler
Copyright © 2019 Pan Stanford Publishing Pte. Ltd.
ISBN 978-981-4800-14-3 (Hardcover), 978-0-429-44856-0 (eBook)
www.panstanford.com

from the environment. The ability of reaction on changes in the environment is dependent on the ability to recognize signals from the environment and to distinguish important from less relevant signals. All reactions, adaptation, and learning have to be based on the receiving and processing of incoming signals.

Classical technical systems with a limited number of sensors are in a dilemma of temporal order: The designer of the system has to predict the expected type and importance of incoming relevant signals before designing the system. Any uncertain future situation and unexpected changes in the environment cannot be considered in the system and sensor design. The classical sensor hardware is not adaptable, and the number of predefined signal input channels is very limited.

Therefore, future sensing concepts need a conversion of the temporal order: The technical system has to be made free from the ultimate need for a predefined signal selection. The design has to be based on the ability for uptake of as many as and as different signals from the environment as possible. This gives the technical system the chance to shift the selection of generation of data out of the flood of incoming signals from the design phase into the operation phase of the system. The technical system will become much more flexible and adaptable by this "late signal selection."

Such a sensing strategy demands for more sophisticated signaling pathways that allow for efficient signal and data processing, including data selection. Each system needs a higher number of very different primary transducers for uptaking signals from the environment and from any operating object, investigated, processed, or manipulated target. These primary transducers supply data that can be further used in the technical system and be distinguished, evaluated, and selected at an early stage. These primary transducers have to be as small and flexible as possible. But their variability and mobility should be high. These requirements give the deciding systemic reason for the particle-like character of these transducers. For designing of sensor particles a large spectrum of particle properties has to be respected (Fig. 2.1). It includes general physical parameters as well as chemical and biological features that are important for the specific application circumstances of the particles.

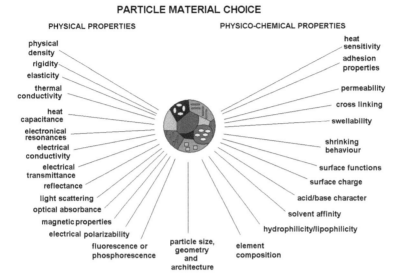

Figure 2.1 Physical and chemical properties that have to be reconsidered for the material's choice for sensor particles.

Besides this general challenge for particle-type transducers, there is a second very important aspect for the development of particle-based sensing systems: A lot of interesting signals are coming from complex, spatially organized objects and demand for a high local resolution of signals. This fact becomes immediately clear when details of processes in biological tissues, cell assemblies, or microbial communities are objects of signal reception. Toxicological and metabolic tests as well as other screening operations demand for a parallel data readout from thousands, millions, or even billions of cells. Ultraminiaturized and individually interacting transducers are required for massive parallelized data transfer from these natural micro-objects into technical systems.

What are the main issues for the development and application of particle-based sensors? There are five basic requirements that have to be addressed by the particles (see Fig. 2.2):

- Recognition
- Signal transduction
- Signal transfer
- Object targeting
- General system compatibility

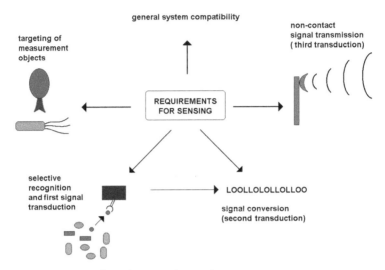

Figure 2.2 General requirements for sensing.

A sensor particle must be adapted to all of these requirements in order to become applicable. This challenge can be satisfied by choosing the right basic material of the particle, functional molecules, or substructures and a suitable general structure of the particle. In addition, the particles must have suitable physical properties such as physical density; a certain electrical conductivity or dielectric behavior; dia-, para-, or ferromagnetic properties; suitable heat capacitance and thermal conductivity; and specific transmittance, scattering, or absorption of different parts of electromagnetic spectrum. Besides the physical properties, some general chemical and physicochemical properties are important, such as surface hydrophilicity or hydrophobicity, acid or base character, anionic or cationic surface charging, matrix cross-linking, solvent affinity, swellability, and shrinking behavior. Finally, the size and geometry of particles are important for their applicability.

The recognition function is the most fundamental requirement for each sensor. The particle must be able to take up the desired information from a target object. The recognition function can be very differently organized depending on the specific information. On the one hand, the particle may react on a signal from outside as whole, for example, by changing a global property. Signal-triggered changes in swelling, in size or shape, and in density or elasticity are typical examples of such a general response. On the other hand, particles can receive signals from outside by single components only. In an extreme case, only one particle-attached or particle-integrated molecule acts as the sensing receiver, for example, by a fluorescence, a chemoluminescence, or a photochemical response.

Solely, recognition is not sufficient for a sensor function. The second important requirement is the efficient conversion of the incoming primary signal into a secondary signal form that can better be processed by the technical system than the primary signal. This conversion is described by the term "sensoric transduction." This function means the reliable offering of a readable signal in any case of reception of a certain input signal. The transduction has to be always to represent a clear causal chain: If a certain input signal appears, then have the sensor generate the complementary secondary output signal. Wrong output signals have to be avoided. A low probability of signal output without an external input signal (low background noise) is required for each reasonable sensor application.

Recent sensor concepts are fed by a convergence between the classical technical and a biomimetic approach. An idealistic biomimetic approach integrates the sensor function into a miniaturized compartment that might be constructed in analogy to cells. It consists of a cell membrane-analog capsule in which bioreceptor-like sensing molecules are integrated and of an interior with molecular functions like biocatalysis and molecular signal conversion. Such objects could be equipped with several recognition sites and several nano- or microscaled transducer systems and have the character of an "intelligent multifunctional particle" (Motornov et al., 2010).

What are the quantitative requirements for particle-based signal transduction? Each transduction is connected with a certain transport and conversion of energy. This need is independent of the type of input and output signal and only determined by the thermodynamic sensitivity and the dynamic range of the sensor. The maximum power of a transducer system is given by the product of the minimum detectable signal energy E_{min}, the minimum time constant (integration time) τ, and the dynamic range factor D, which characterizes the measurement range and the accuracy of signal processing. The values for the sensor power can be very different depending on the sensor type. Some examples are given in Table 2.1.

$$P_{max} = E_{min} \times D/\tau \tag{2.1}$$

Table 2.1 Signal power in relation to signal energy and addressed dynamic range of sensor signals

Minimum signal energy	Dynamic range (bit)	Dynamic range factor (approx.)	Time constant	Maximum power
1 fJ	8	256	1 s	0.256 pW
1 pJ	10	1000	0.1 s	10 nW
1 nJ	12	4 000	1 s	4 µW
1 nJ	16	64 000	0.1 s	640 µW
1 µJ	18	256 000	10 ms	2.56 W

One microjoule is a rather comfortable energy in recent sensing. It corresponds to a temperature difference of 1 K for a 0.24 nL droplet (0.00024 µL) of water approximately. Measurements at the nanojoule or even the picojoule level are much more challenging and demand for the detection of temperature differences at the milli-Kelvin and the micro-Kelvin level, respectively, for the same small volume. The first line in the table means to step further downward on the energy staircase for 3 orders of magnitude. This step leads into a complete new class of requirements, the detection of single-particle events: 1 fJ is the order of magnitude of the kinetic energy of a single particle with one elementary charge, if it was accelerated at about 10 keV. This energy is very common for single particles in technical vacuum processes, for example, in the fabrication of semiconductor devices. Single photons of X-rays can have energies of the same order of magnitude.

Reconsidering typical measurement—or integration—times each of the mentioned energies can be directly used for an estimation of processed power of the signal transduction. Sensing in the megawatt region is more or less trivial when conventional sensors are used. But what are the consequences if microparticle-based sensors are applied for the transduction of energies in this order of magnitude? The estimation of the power density is a suitable strategy to evaluate the applicability of small sensor particles. The examples in the Table 2.2 illustrate this aspect.

Table 2.2 Illustration of the effect of power density in small sensor particles by comparison of related time constants of a thermal process in a conventional system (water boiling)

Power of signal transduction	Particle size (approx.)	Particle volume	Power density	Approximated boiling time for a 1 L flask at the same power density
0.256 pW	0.1 mm	1 nL	0.256 mW/L	1.2 Gs/~3 × 10^5 h
10 nW	0.1 mm	1 nL	10 W/L	32 000s/~9 h
4 μW	1 mm	1 μL	4 W/L	80 000 s/~73 h
4 μW	0.1 mm	1 nL	4 kW/L	80 s/~1.3 min
640 μW	0.1 mm	1 nL	0.64 MW/L	0.2 s
2.56 W	1 mm	1 μL	2.56 MW/L	50 ms
2.56 W	0.1 mm	1 nL	2.56 GW/L	50 μs
(80 kcal = 320 kJ)				

The high power of signal transduction means an enormous energetic stress for particles of small size. In particular, for large, dynamic ranges, particle-based sensing technologies have to combine high sensitivity with high tolerance against energetic stress.

2.2 Information Transport by Particles

Particles can act as carriers of information if they integrate transportability properties with the ability of uptaking, storage, and release of information. A usual example for this function is represented by labeling particles. This class of particles must allow, at least, the implementation of a data set, its safe storage, and the compatibility with a readout mechanism. Microbarcode particles represent a typical class of such objects.

The properties and manufacturing of barcode particles are mainly determined by the chosen readout procedure. Optical imaging is favored for that because it can be performed without direct contact, can work fast, and is applicable on high information transfer rates. The availability of cheap and powerful optical microdevices and micro-optics as well as the widespread use of consumer devices like smartphones integrating powerful cameras support an increasing number of imaging-based readout of microbarcode labels.

An optical readout can be based on spatial information, on a spectral coding, or on a combination of both. The first corresponds to the classical barcoding with black-and-white strips. This strategy means a higher effort for the implementation of the spatial contrast, which demands well-defined internal particle structures but means, in general, a lower challenge for the readout by imaging. The second can be realized by unstructured particles. The coding is then based on a mixture of chromophores. Coding information can be implemented by a combination of different spectral channels (colors) and different color depths (absorbances). In Section 5.8., some examples of particle barcodings are given.

The capacity of stored information depends on the number of distinguishable spectral channels and the resolution in absorbance (Fig. 2.3). The comparatively large bandwidth of optical absorbances in condensed matter does not allow for a high number of optical channels if only the visible range is used, on the one hand. On the other hand, high absorbances of many substances in the ultraviolet (UV) range and the high and broad absorbance of water and aqueous solutions in the infrared (IR) range led to a preferential use of the visible part of the optical spectrum. In addition, the high availability of powerful miniaturized optical devices such as light-emitting diodes (LEDs), laser diodes, and charge-coupled device (CCD) chips cause a focusing of the visible light and its transition regions to the near-infrared (NIR) and near-UV region.

In most cases, four spectral channels can easily be addressed separately. A purely qualitative combination would allow us to distinguish 16 different particle types (0, A, B, C, D, AB, AC, AD, BC, BD, CD, ABC, ABD, ACD, BCD, and ABCD). A much higher number of types can be addressed if the color content is quantified and can be measured. Sixteen different gray levels (4-bit level) mean 65,536 distinguishable colors if 4 color channels can be used. If only 3 channels (RGB, for example) can be used, this number is reduced to 4096. Besides classical preparation technologies, microfluidic approaches offer a convenient way for generating polymer microparticles with a very narrow size distribution (Serra and Chang, 2008). Computer-controlled pumping of different components allows for an automated generation of differently dyed particles (Visaveliya and Köhler, 2015).

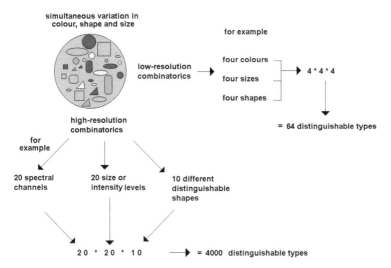

Figure 2.3 Variability of labeling composite particles by combination of embedding small particles with a different size, shape, and color.

The storable information is strongly dependent on the accuracy of controlling the color content and the size of particles, on the one hand, and the precision of photometric measurement, on the other hand. A rough estimation of this problem can be deviated from Lambert–Beers law for the absorbance E, where ε is the molar absorbance (extinction) coefficient, c is the (molar) dye concentration, and d is the optical path length (approximated by the diameter of the dyed particles):

$$E = \varepsilon \times c \times d \tag{2.2}$$

It describes the ratio of transmitted light, I, and incident light intensity, I_0:

$$E = -\log(I/I_0) \tag{2.3}$$

For labeling with polymer microparticles, it is easy to realize high absorbances because a lot of organic dyes with molar extinction coefficients between 10,000 and 100,000 L/(mol.cm) are available and can be applied in high concentrations in the polymer matrix of the labeling particles. Table 2.3 gives some examples of the maximum absorbances E_{max} for moderate dye concentrations.

Table 2.3 Maximum absorbances depending on particle sizes and extinction coefficients estimated by Lambert–Beers law

Optical path length (approx. particle diameter)	Matrix concentration of dye	Molar extinction coefficient	Maximum absorbance	I/I_0
0.5 mm	2 mM	20,000	2	0.01
0.5 mm	1 mM	20,000	1	0.1
0.5 mm	1 mM	10,000	0.5	0.32
0.1 mm	15 mM	15,000	2.25	0.0056
0.1 mm	4 mM	20,000	0.8	0.16
50 μm	4 mM	20,000	0.4	0.4
50 μm	20 mM	15,000	1.5	0.032
20 μm	20 mM	15,000	0.6	0.25
5 μm	20 mM	15,000	0.15	0.71

The fabrication of microbarcode particles is possible by different methods. Higher numbers of strips with different widths can be manufactured by a combination of film deposition followed by anisotropic trench etching and releasing of particles from the surface by isotropic etching of a sacrificial layer. Sputtering, vapor deposition, spinning, and galvanic film deposition techniques are suitable for generating film stacks for nanobarcodes. But the readout of strips with widths in the submicron or the nanometer range requires high-resolution optical techniques or even ultramicroscopy, which is not well applicable for a broader use.

A simpler way for enhancing the addressable space is the application of composed microparticles. Their size can be adapted to the requirements of the labeling process and the readout devices. The potential of particles composed by a small set of dye-labeled smaller particles is also illustrated by a simple calculation: If three particles of optically distinguishable size and the above-mentioned 65,536 different colors are merged together, the resulting total number of addressable classes is $(65,536)^3$, which is about 2.8×10^{14}, a really huge number.

But it can be clearly said that simple microbarcode particles do not have a true sensor function. They can transport and store information. But they are passive objects. They don't have any function for signal conversion. Thus, they can support information storage and identification of objects, but they cannot be regarded as "mobile spies."

2.3 Signal Conversion by Particles

A sensor converts one type of signal into another type. The typical case for this conversion is the generation of an external signal into a signal of a system-inherent general communication system. In most recent technical devices, digital electronic communication forms the technical basis for system-inherent communication. This means that all forms of external signals have to convert into digital electronic signals. Depending on the character of the external signal, two or more steps of signal conversion can be required for generating a related digital electronic signal finally. In most cases, the primary signal is converted, first, into an analogous electronic signal, which is, in the following step, converted into a digital signal:

Extern signal → Primary sensing → ... → Analogous electrical signal → Analog/digital (AD) conversion → Digital electronic signal

In principle, sensor particles could take tasks in each step of this signal conversion chain. But normally, powerful chip-based AD converters do this job. This last step can easily be standardized. AD conversion has the character of a universal operation, which is independent of specific measurements or the specific operating system. Thus, one and the same AD converter device can be part of very different transduction chains (Fig. 2.4).

The opposite situation is found in the first steps of the transduction chain. Here, incoming signals of a very different character have to be received and converted. Specific receiver and converter elements are required for the sensing. Many signals—among them particular mechanical activity, viscosity or hardness measurements, and tactile sensing tasks—can only be received with direct contact between the investigated object and the primary transducer. Direct mechanical contact is also required in many cases of chemical and biochemical sensing. In these cases, particle transducers with specific receiver functions can act as primary signal converters.

typical signal transduction chain of a molecule-specific
sensor particle with secondary optical signal transfer:

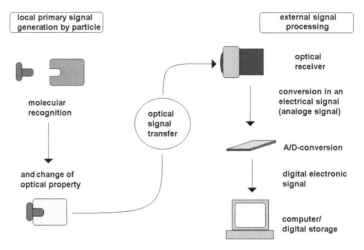

Figure 2.4 Typical transduction chain from the particle-based primary recognition to the final sensor signal.

Despite autonomous behavior of particle-like micro-robots (see Chapter 6), it is hardly to imagine that the integration of a complete conversion chain up to the digital electronic signal would represent an efficient solution for sensor particles. Much more usual is the conversion of the specific primary signal into another more system-compatible signal form, which can be further transmitted from the sensor particles into the electronic system. In most cases, this system-internal signal conversion and transmission proceeds between particles and fixed transducers, which means between mobile primary transducers and a stationary secondary transducer. A noncontact signal transfer is the best solution for this task.

2.4   Particles as Primary Transducers and Secondary Noncontact Signal Transfer

Sensor particles are always favored if primary direct contact sensing is required. Then, sensor particles have to combine the following functions:

Receiving of external signal by direct contact → Conversion into a secondary signal → Noncontact transmission of the secondary signal to a secondary sensor and an information processing system

The first two steps have to be active properties of the sensor particle. The third step can be realized by an active or even by a passive function. Instead of an active release of signals, the primary received and converted information can also be transmitted by a change in a suitable property of the particle, which can be read out by a general mechanism in the transducing system. This difference can easily be illustrated by the difference between the signal conversion by chemoluminescence-based sensing and by sensing using a simple color change. In the first case, a chemical recognition proceeds first by a chemical reaction on a sensor particle. This reaction is coupled with the chemical formation of a product molecule in an electronically excited state. This state will relax by releasing a photon into the environment, which can be detected by a universal light receiver system. The chemoluminescent primary sensor can be understood as an active transmitting system because it sends out the photon only in the case of the ongoing primary recognition reaction. In the second case, the primary chemical recognition reaction on the particle causes a color change, which can only be recognized if the sensor particle is illuminated. Then the light-sensitive secondary receiver system can distinguish between differently colored or higher- or lower-pigmented particles.

Optical channels are mainly used for secondary noncontact signal transfer from the primary sensing particles into a secondary sensor and information processing system. These optical channels can vary in their properties by different geometries in the optical system, by different intensities and energies, and by different spectral ranges.

The central requirement for sensor particles for noncontact signal transmission to a secondary receiver is the ability to change a readable property. In addition, this change must never proceed spontaneously, but it should have to be exclusively induced by the primary incoming signal.

The spectrum of changeable properties of microparticles is large. It includes geometrical features (size and shape), mechanical features as external and internal motion, elasticity, chemical composition, surface states, temperature, electrical and thermal conductivity, electrical charging, and electronic excitation. For a repeated sensor activity, the particle state after the decline of the primary signal and the transmission of the secondary signal should switch into the original particle state by a relaxation process. The following table

(Table 2.4) gives some examples of sensor particle properties that can be used for noncontact secondary signal transmission.

Table 2.4 Particle properties usable in functions for signal transduction

Changeable particle state	Affected by primary external signal (examples)	Response mechanism (examples)	Noncontact signal
Size	Swelling	Volume increase	Imaging, DLS
Shape	Local swelling	Local volume increase	Imaging
Shape	Temperature change	Bicomponental expansion	Imaging
Shear stress	Stretching, orientation	Change of shape and/or size	Imaging, light polarization
External motion	Mechanical force	Acceleration, sedimentation	Imaging, electrical conductivity
Internal motion	Mechanical force	Vibrational resonance	Imaging, DLS
Elasticity	Mechanical force	Deformation	Imaging
Chemical composition	Reactive species	Chemical reaction	Optical absorption, fluorescence
Color	Light	Photochemical reaction	Imaging, photometry
Surface state	Adsorption	Interface affinity	Reflectivity, electrical conductivity
Electrical charging	Electrochemical reaction	Electrophoretic effect	Imaging, electrical conductivity
Temperature	Heating, cooling	Color change	Imaging, photometry
Thermal conductivity	Heating, cooling	Kinetics of color change	Time-resolved imaging
Electronic state	Activation substrate	Chemical activation of excitation	Chemoluminescence
Electronic state	Quencher interaction	Fluorescence quantum yield	Fluorescence intensity

DLS, dynamic light scattering.

The following six important issues have to be addressed by the design of sensor particles:

- Sensitivity
- Specificity
- Accuracy
- Response time
- Reversibility
- Robustness

For high sensitivity, the sensor particles have to react on low input signals. High sensitivity in chemical sensing means either a significant reaction on changes in concentration or even the interaction with a low number of molecules. A high sensitivity in chemical sensing can be expected in the case of very small particles because in this case a small number of reacting molecules or even a single molecule can change physical properties of the sensor particle significantly. But such high sensitivity can only be used if the secondary signal transducing system is able to read out the change of the properties of small particles.

The demand of accuracy is related to the quantification of sensing. This means that a sensor particle has to react on an incoming signal by a gradual change of its properties. This gradual change has to be recognized and quantitatively differentiated by the secondary signal transfer in order to keep the information and to realize a quantitative sensing. In particle-based sensing, the finally formed sensor signal is always a superposition of the response function of the primary transduction by the particle and the response function of the second signal transfer from the particle to the final detection system.

Reversibility is one of the most critical aspects in sensing and in particular in chemical and biochemical sensing. It is an important challenge in particle-based sensing, too. A well-working sensor particle has to come back to its original state after the primary signal disappears. A certain time needed for the relaxation process has to be taken into account for the definition of measurement frequencies. The minimal measurement time is determined by the sum of the response time itself and the required relaxation time.

Reversibility in chemical sensing can easily be achieved if the sensor bead reacts by a fast relaxing chemical equilibrium on the change in the outer conditions. Protonation/deprotonation equilibria are a typical case for such fast-relaxing chemical systems. If parts of a chromophore system are protonated or deprotonated, the sensor particle changes its color. Such proton-sensitive particles are useful for pH sensing. Besides the resonance frequency in absorption, the quantum yield of fluorescence dyes is also frequently affected by pH. Thus, sensor particles for pH measurement can also be based on suitable immobilized fluorescence dyes.

Chapter 3

Optical Solutions: Synthesis and Applications of Optical Sensor Particles

3.1 Optical Transduction Principles for Analyses and Sensing

The change of color due to the interaction of an indicator dye with an analyte is one of the simplest possibilities of signal transduction with optical readout. It is largely used in traditional lab applications as well as in recent automated analytical devices. But, often the measurement of concentration by optical absorption suffers from the low change in the absorbance or by low specificity and by interferences with other components of the investigated medium. In particular in the case of low concentrations, in the case of low extinction coefficients, and in the case of measurements in thin films or adsorbate layers, the differences between incident and transmitted light are too low for a reliable measurement. In these cases, fluorescence measurements are much better suited for detection.

Fluorescent sensor particles can either be generated by incorporation of fluorophores into a permeable particle matrix or by immobilization of fluorophores onto the particle surface. The immobilization on the surface has the advantage of free access by analyte molecules that can interact with the fluorophores directly. The

Mobile Microspies: Particles for Sensing and Communication
Michael Köhler
Copyright © 2019 Pan Stanford Publishing Pte. Ltd.
ISBN 978-981-4800-14-3 (Hardcover), 978-0-429-44856-0 (eBook)
www.panstanford.com

drawback of surface functionalization is the lower total fluorophore concentration that can be achieved. Matrix incorporation allows us to realize a much higher total fluorophore content. But, the access of analyte molecules to the fluorophores is limited by the permeability of the matrix material.

Fluorescence sensing is applied for chemical characterization. It is based on changes in fluorescence signals by interaction of a fluorescent component with analyte molecules. The related sensor particles can be characterized by the following general process chain:

Molecular interaction → Change of fluorescence properties → Absorption of measurement (excitation) light → Emission of fluorescence light

There are two principles in the change of analyte-dependent fluorescence signals: The first is the change in fluorescence quantum yield without significant change in the optical spectra. This effect can be due to a change of molecular mobility of the fluorophore and the increase or decrease of rate of radiation-less electronic des-activation by thermal relaxation. Or it can be due to a transfer of excitation energy from the excited fluorophore to the analyte molecule, accompanied by thermal dissipation of energy. In both cases, the resulting quantum yield is dependent on the analyte concentration and can be used, therefore, for the characterization of analyte concentrations. The second principle is the shift of the resonance wavelength of fluorophores due to the interaction with the analyte. Such an effect can be caused by specific chemical reactions and changes in the chromophore system. It can be read out either by a change in the fluorescence quantum yield, which can be due to a shift in the absorption maximum. Or it can be marked by a significant change in the Stokes shift, which means the distance between the absorption maximum and the fluorescence emission maximum at the wavelength axis.

In comparison with absorption measurements, fluorescence sensing has always the enormous advantage of much higher sensitivity (Fig. 3.1). This is mainly due to the fundamental difference in the formation of the signal-to-noise ratio in both types of measurements. Absorption measurements are always based on the comparison between illumination light and transmitted (or reflected or remitted) light of a sample. It is in the nature of this method that both intensities have to measure at the same wavelength. At low chromophore concentration, the difference between incoming light (I_0) and transmitted light (I) becomes very small, and their ratio is very close to 1. This means that a very small intensity difference has to be detected on high background intensity. Small deviations in the intensity of light source, small variations in the optical pathway, or other small disturbances can cause strong deviations in the measurement result. The consequences are low signal-to-noise ratios in the case of low analyte concentrations. The situation in fluorescence measurements is completely different. Fluorescence needs absorption of photons, but the non-absorbed photons are not the primary objects of detection but light of a different wavelength. Fluorescence is a phenomenon of light-induced light emission. That means that it represents a photoluminescence method. Due to the Stokes shift, the emitted light can always be measured at a wavelength that is at a certain distance from the excitation wavelength. Thus, a spectral separation can be realized using optical filters or monochromators. Typically, a steep high-pass filter is used in the excitation beam in order to exclude all higher wavelengths from the light source. Complementarily, a low-pass filter is applied in the fluorescence channel in order to suppress any residual parts of the excitation light. In addition to the spectral separation, it is possible to improve the signal-to-noise ratios by a geometrical separation, too. Therefore the fact that the emission of fluorescence light occurs in all space directions is used. Thus, a geometrical separation can simply be achieved by an orthogonal arrangement of the excitation beam and the emission beam. In the overwhelming part or applications these both strategies allow us to detect very low intensities of fluorescence and, therefore, the detection of very small analyte concentrations, sometimes down to the single-molecule level. This high sensitivity is also an important reason for the applicability of fluorescence detection in sensing using very small sensor particles.

Fluorescence offers, in addition, a third possibility of signal separation. This possibility is related to the lifetime of the fluorophore in the electronically excited state. This lifetime is typically in the nanosecond range. This is too short for separation by mechanical beam chopper, but it can be realized, recently, by fast-switching electronic shutters. In these measurement arrangements, the fluorescence channels remain closed during the electronic excitation, for example, by a nanosecond flash. It is only opened after the decay of the excitation flash or the closing of the excitation channel. This temporal separation can be used in combination with the spectral and the geometric separation of excitation and emission light and lead to very high sensitivities in fluorescence measurements.

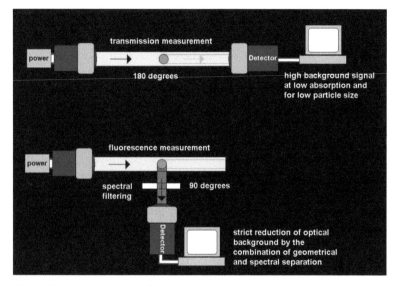

Figure 3.1 Comparison of principle arrangements for transmission and fluorescence measurements.

The temporal separation is still easier to realize in the case of higher excitation lifetimes. High lifetimes with typical values in the order of magnitude of 1 ms—that is, the excited state lives about a million times longer than in the case of a normal fluorescence— are typical for the phenomenon of phosphorescence. In analogy to fluorescence, the emission of phosphorescence light is also caused by a previous absorption of excitation light. The deciding difference is that fluorescence is going on without any change in the spin multiplicity of the chromophore system during the excitation. In contrast, a spontaneous spin switching is observed during the transition from the primary formed excited singlet state into a secondary formed excited triplet state. The parallelization of two electron spins leads to a little lower total energy of the system. This effect causes a further reduction of the energy of photons that are emitted during the electronic deactivation from the triplet state. An important consequence for optical sensing is the increase in Stokes shift, which simplifies the spectral separation for excitation and emission light in the case of phosphorescence measurements. The second effect is the much lower probability of spontaneous transition of the excited triplet system into the ground-state singlet system, which demands for a spin conversion. This low transition probability is responsible for the high phosphorescence lifetime in comparison to a typical fluorescence lifetime.

All the mentioned general advantages of photoluminescence detection can be exploited by particle-based sensing. But, in contrast to the simple application of fluorescent indicator dyes, the particle-based measurements have four additional important advantages: First, the dye is always in a certain chemical and physical microenvironment that is dominated by the particle matrix and is only moderately affected by other components in the outside liquid. Second, it is separated by any attack from cells or microorganisms that are two large for passing the pores of a swollen polymer or even a gel matrix and cannot penetrate the particle material. In consequence, the fluorophores and other molecular structures, which are used in the particle-based sensing, are protected against metabolization by microorganisms or cells. And third, the particle matrix can also be used for controlling the permeation of different analyte components. Thus, the construction of the matrix, their pore sizes and internal molecular mobility, their functional groups, and their electrical charges can be varied in order to control the molecular transport and to improve the selectivity of particle-based sensing.

3.2 Optical Sensor Particle Types

3.2.1 Fluorescence Sensing

The application of fluorescence sensor particles gives a unique opportunity for intrinsic referencing of optical signals if an imaging procedure is applied. They ensure always a local contrast. This is the opposite situation as in the case of molecular dispersed indicator molecules. The spatial resolution of imaging techniques can easily be scaled down to the lower micrometer level. Meanwhile, the high availability of low-cost but high-performance optical array detectors and miniaturized optics supports the application of photoluminescent sensor particles and its use in spatial-resolved analytical measurements.

A simple way is to concentrate the fluorescence dye molecules in a suitable carrier particle. The fluorophores can either be immobilized in a film at a particle surface or be distributed inside the particle matrix or a special part of a composed particle (Fig. 3.2). Many organic fluorescence dyes, as well as fluorescent nanoparticles (NPs), can be dispersed and immobilized inside polymer micro- or nanoparticles.

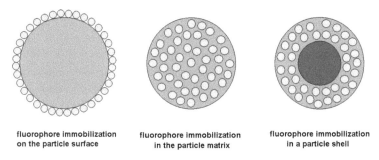

fluorophore immobilization
on the particle surface

fluorophore immobilization
in the particle matrix

fluorophore immobilization
in a particle shell

Figure 3.2 Constructing fluorescent micro- or nanoparticles by immobilization of fluorophores on the particle surface by incorporation inside the particle matrix or by immobilization inside the shell of a core/shell particle.

Recently, a large spectrum of fluorescent inorganic as well as organic nanoparticles (NPs) has been discovered. Among the inorganic particles, compound semiconductor particles (quantum dots [QDs]) and rare-earth-doped particles, for example, europium- and ytterbium-doped sodium/yttrium fluoride NPs (Wang and Li, 2006) and CdSe QDs (Gao et al., 2002), are of interest for labeling and sensing applications.

Metal NPs with a larger number of atoms are not suitable for fluorescence sensing because they normally cause a quenching of excitation states. This situation is changed only if the metal particles are small and have the character of clusters. It was found that silver nanoclusters of less than 100 atoms show their own fluorescence

(Diez and Ras et al., 2011), which make them very interesting for biomedical sensing applications (Lin et al., 2009). It was found that the fluorescence intensity is significantly dependent on the molecular environment or the conjugation partners of the silver nanoclusters. Thus, such clusters carrying single-stranded DNA had been found to be able to distinguish between hybridized and nonhybridized states, which can be used for proving matching sequences of micro-RNA (Yang and Vosch, 2011).

A typical example of fluorescence sensing by microparticles is realized in swellable polymer microparticles for pH measurements (Nagl and Wolfbeis, 2007). In such particles, a fluorescence dye with a strong pH-dependent fluorescence quantum yield is used. The dye is immobilized in a polymer matrix that can swell in a suitable solvent, for example, water. In the swollen state, protons from the environment can easily diffuse into the matrix and interact with the dye molecules. The application of such sensor particles is of particular interest for the evaluation of local pH in biological systems as droplets of cultivation liquids, in biofilms, or inside living cells (Fig. 3.3). pH detection has a big advantage in comparison to the detection of other analytes that protons are very small particles and having high permeation ability. This means comparatively low requirements for the swellability and porosity of the polymer particle matrix.

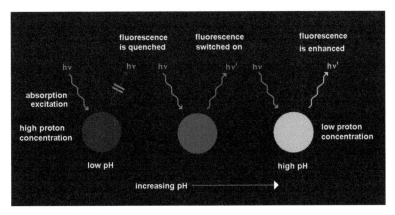

Figure 3.3 Measuring pH by fluorescent nanoparticles responding by pH-dependent fluorescent quantum yields.

Table 3.1 Examples of fluorescent nanoparticles

Material	Shape and size	Function	Ref.
Amphiphilic block copolymers	30 nm	Ratiometric HOCl imaging	Wang et al., 2017
BSA, cross-linked and dye-doped	35 nm	pH sensing	Stromer and Kumar, 2017
Carbon NPs, functionalized	<10 nm	Labeling for cell imaging	Bhunia et al., 2014
CdSe, ZnS-capped	2–6 nm	pH sensing	Gao et al., 2002
CdSe/ZnS, core/shell, spiropyran-modified	Sphere, about 6 nm	Light-controlled switching	Zhu et al., 2005
Curcumin-functionalized silica NPs	25–35 nm silica core, 4–8 nm organic shell	Cholesterol sensing	Chebi et al., 2017
Dipeptide NPs, functionalized		Biomolecule detection	Fan et al. 2016
Dye-doped silica NPs	Between ~5 and several 100 nm	Selective recognition, labeling	Schulz and McDonagh, 2012
$Na(Y_{1.5}Na_{0.5})F_6$ rare earth-doped	Submicron rods		Wang and Li, 2006
Organic nanocapsules with AIE-dyes	~10 nm	Labeling for cell imaging	Xu et al., 2016
Polymer NPs, imprinted	30–60 nm	Detection of cancer cells	Liu et al., 2017
Polythiophene NPs	~50 nm	Targeting of lysosomes	Zhao et al., 2017
Platinum-complex-doped PS particles		Oxygen sensing	Zhang et al., 2016
Silica NPs with FRET systems	70 nm	Labeling, barcoding	Wang and Tan, 2006
Surface-mod. cationic polyelectrolyte polymer	Spheres, ~0.5 μm	Detection of enzyme activity	Chemburu et al., 2008

Instead of pH-sensitive organic dyes, inorganic NPs can also be applied for pH sensing. This method was applied, for example, by use of N-rich carbon NPs (Shi et al., 2016). These particles are marked by low cytotoxicity and could be applied for imaging of pH of living T24 cells.

In some cases, the interaction of metal ions with ligands can result in enhancement in fluorescence quantum yield. This effect is used, for example, in sensor NPs conjugated with far-red-fluorescent dyes, which show a strong enhancement of fluorescence in the presence of aluminum ions and can be used, therefore, as Al sensor particles (H. Liu et al., 2013). An increase of fluorescence was also observed in fluorescent cyclodextrine-conjugated polymer NPs (Xu et al., 2012). The cyclodextrine unit is responsible for the host character of the sensor particles. The fluorescence is significantly enhanced if surfactants such as sodium dodecyl sulfate (SDS) or sodium dodecylbenzene sulfonate (SDBS) are incorporated.

The measured intensity of fluorescence is related to the kinetics of electronic deactivation of the excited state. Time-resolved fluorescence measurements supply the decay times. In the case of significant differences, the measurement of fluorescence decay can allow us to detect two different relaxation processes in parallel in order to realize simultaneous determination of two biological or biomolecular targets. This concept was successfully applied by Hoffmann et al. (2013) for the detection of fibroblasts and macrophage cells using fluorescent NPs with an emission in the near-infrared (NIR) region.

In contrast to the measurement of analytes by change of fluorescence intensity, the binding probability can also be used for signal transduction via fluorescence. In this case, NPs are advantageously used for realizing high fluorescence or phosphorescence intensities, but analyte binding and selectivity are addressed by a chemical or biomolecular functionalization of the particle surface. Fluorescent silica NPs (Schulz and McDonagh, 2012) are a typical example of these special labeling particles (see also Section 5.2.2).

Table 3.2 Examples of nanoparticles for optical detection of metal ion concentrations

Metal ion	Nanoparticle/Ligands	Transduction principle	Ref.
Al^{3+}	N,N'-ethylenbis(salicylimine)	Fluorescence	Huerta-Aguilar et al., 2015
Ca^{2+}	Quantum dot/ Rhodamine-BAPTA-FRET-system	Fluorescence/FRET	Zamaleeva et al., 2015
Cu^{2+}	Porous silica-coated CdSe/ZnS NPs	Fluorescence quenching	Sung and Lo, 2012
Fe^{3+}	Carbon NPs	Fluorescence quenching	Singh and Mishra, 2016
Fe^{3+}	Carboxyl-functionalized carbonnitride NPs	Photoluminescence	Shirivand et al., 2017
Hg^{2+}	Carboxyl-functionalized carbonnitride NPs	Photoluminescence	Shirivand et al., 2017
Sn^{2+}	Salicylaldehyd-modified organic NPs	Fluorescence enhancement	Patil et al., 2017
Zn^{2+}	N,N'-ethylenbis(salicylimine)	Fluorescence	Huerta-Aguilar et al., 2015

The application of selectively reacting chelate ligands forming fluorescent coordination compounds is a general strategy for particle-based measurement of the concentration of metal ions. It is particular suitable for the detection of transition metals that are forming stable complexes and can be distinguished by their ion radii and electrical charge. NPs of *N,N'*-ethylenbis(salicylimine) can be used for the detection of zinc ions (Huerta-Aguilar et al., 2015). In addition, aluminum ions can be detected by fluorescence enhancement due to a co-coordination with zinc chelates.

3.2.2 Phosphorescence Sensing

The phenomenon of phosphorescence is caused by a spin inversion and the formation of a triplet state in the electronically excited state. It is frequently observed in dyes containing heavier elements

(Fig. 3.4). The triplet state is much more stable than the singlet state, which results in strong prolongation of the lifetime of the excited state. But it can easily be quenched by triplet molecules.

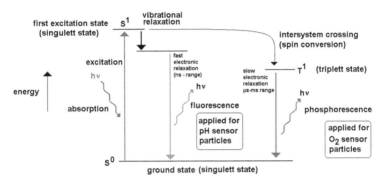

Figure 3.4 Energy scheme for electronic excitation, fluorescence, and phosphorescence for optical sensing (simplified Jablonski diagram).

Phosphorescence-related sensor microparticles are very interesting for analyte particles with unpaired electrons. In particular, the detection of molecular oxygen is possible conveniently because oxygen represents a triplet molecule in the ground state and acts, therefore, as a very efficient quencher for excited triplet states of dyes. Powerful sensor microparticles can be realized, therefore, by incorporating triplet dyes into the polymer matrix of microbeads. Their phosphorescence signal increases with decreasing oxygen concentration due to the reduction of the quenching effect. Such particles are well suited for the monitoring of oxygen consumption in cell cultivation, in particular in microdroplets and for investigation of physiological activity in aerobic metabolism. Small particles of this type can also report the local oxygen concentration in living cells after incorporation in the cytoplasm. Oxygen sensor particles for optical readout have been developed in different configurations: Guice et al. (2005) immobilized tris(diphenylphenanthroline) ruthenium in a layer-by-layer (LBL) shell layer on polymer NPs for obtaining oxygen-responsive sensor particles (Fig. 3.5).

Funfak et al. (2009) applied pH-sensitive microparticles, and Cao et al. (2015) used oxygen-sensitive microparticles for the monitoring of physiological status of bacteria in microfluid segments (Fig. 3.6). These measurements proved to be particularly suitable for the characterization of the concentration-dependent effect of toxic additives as heavy metal ions or antibiotics on the growth and metabolic activity of microorganisms in microfluidic systems. The oxygen consumption–related phosphorescence yield could be used for the comparison with the effector-dependent autofluorescence and photometric signal by an optical multichannel arrangement. Horka et al. (2016) used oxygen-sensitive particles for the monitoring of *E. coli* in microfluid segments. They had been able to calibrate the absolute oxygen concentration and demonstrated the measurement of the highest oxygen consumption during the growth phase with the highest growth rate.

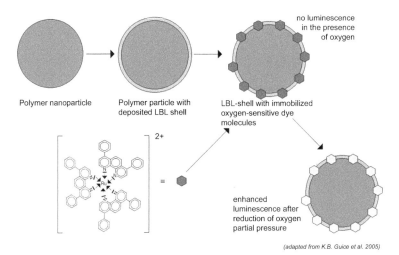

Figure 3.5 Construction of LBL core/shell particles containing an oxygen-sensitive ruthenium complex for particle-based oxygen measurement (adapted from Guice et al. 2005).

The local distribution of oxygen in tissues is particularly interesting for studying the function and pathogenic states in neuronal cells and other components of the brain. Dimitriev et al. (2015a, 2015b) applied oxygen-sensitive phosphorescence particles for investigating tissue oxygenation and metabolic activities in different tissues, among them mouse embryonic fibroblasts, rat pheochromocytoma (PC12), mouse primary cortical (E16), and cerebrellar granule neurons (P7). The authors emphasized the importance of the developed method for future applications in drug delivery systems, particularly for monitoring neural tissues.

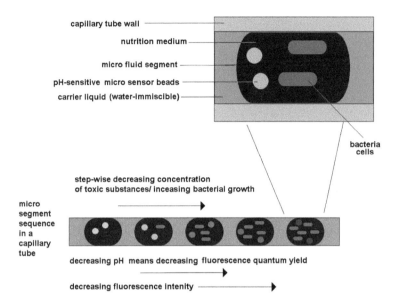

capillary tube wall

nutrition medium

micro fluid segment

pH-sensitive micro sensor beads

carrier liquid (water-immiscible)

bacteria cells

step-wise decreasing concentration of toxic substances/ inceasing bacterial growth

micro segment sequence in a capillary tube

decreasing pH means decreasing fluorescence quantum yield

decreasing fluorescence intenity

Figure 3.6 Application of pH-responsive fluorescent microparticles for the detection of physiological activity of bacteria and for characterization of dose/response functions for toxic substances on growing cell cultures in microfluid segments.

Phosphorescent iridium complexes had been incorporated in silica NPs and used for the detection of hypochlorite. The determination of the comparatively small changes of fluorescence intensity with the analyte concentration was supported by the development of core/shell particles, including an analyte-sensitive shell and a nonsensitive core with a different emission wavelength for referencing.

Besides the direct interaction of analyte components with sensing chromophores, particle sensing can also be applied for more sophisticated labeling, signaling, and measurement strategies. Besides photoluminescent molecules, even fluorescent NPs can be used in such labeling procedures. Compound semiconductor NPs supply high fluorescence intensities. They are available in high variability for different wavelengths. Resonance energies can be tuned by the size of such so-called QDs. Unfortunately, many of these particles are corrosion sensitive and release toxic ions when they are chemically attacked. Therefore, their applicability for monitoring of living systems is limited. Despite this problem, they are under investigation for quenching and switch-on strategies in particle-based sensing (see later) because of their high brightness.

3.2.3 Signal Formation by Molecular Fluorescence Quenching

Sensing by fluorescence quenching can easily performed if the analyte acts more or less specifically as a quencher for the fluorescence of molecules that can be attached to the surface of a sensor particle or are embedded in a permeable particle matrix. The typical sensor particle is formed by a swellable polymer or a gel matrix, in which the analyte-sensitive fluorescence dye is embedded. The trick is to find dyes that are highly fluorescing and can be quenched efficiently and selectively by the targeted analyte. Qian et al. (2015) used NPs formed by self-aggregation of a fluorescent amphiphilic polymer. The transduction results from the interaction of the fluorophores with analyte dopamine. An alternative to quenching is the signal transfer to an excitation energy acceptor by fluorescence resonance energy transfer (FRET) that is able to emit light. There are several particle-based strategies for FRET sensing and combinations of FRET emission and fluorescence quenching (see Section 3.2.10).

3.2.4 Signal Transduction by Nanoparticle-Caused Fluorescence Quenching

In general, metal NPs are efficient quenchers for molecular fluorescence. The excitation energy is quickly transferred from the excited state of the molecule to the metal after the absorption of a photon by a molecular chromophore, normally. Then, the energy is dissipated in heat by a thermal relaxation process. The energy transfer from the molecule to the metal means a strong reduction of the lifetime of the excited state of the molecule and a drastic reduction of fluorescence quantum yield.

The energy transfer is dependent on the distance between the chromophore and the metal. In diluted solutions, the distances are large, which results in non-efficient energy transfer and high fluorescence quantum yields. In the case of a direct bonding of the chromophore to the metal NP the transfer is strongly enhanced. A selective analytical signal can be generated if molecular recognition groups for a certain analyte are immobilized at the surface of a metal NP. This recognition groups bind the analyte and link it to the metal. The fluorescence is immediately suppressed if the analyte fluorescence itself is used for the signal transduction. A second binding or labeling of the analyte with a suitable fluorescence dye is required in the case that the analyte has no suitable own fluorescence activity.

Normally, noble metal NPs are required for this strategy of particle-based detection. The metal particles have to be robust against oxidation. A nonpassivated pure metal surface is required, or at least advantageous, for the fluorescence quenching. Gold is particularly suitable because of its high chemical stability and the convenient possibility of immobilization of molecular recognition by binding versus thiol groups.

3.2.5 Signal Transduction by Nanoparticle-Enhanced Fluorescence

Despite the fact that metals are quenching molecular fluorescence in most cases (previous section), there is the possibility of a fluorescence enhancement if the NPs themselves can be excited efficiently by light and transfer their energy onto adsorbed molecular fluorophores.

The electromagnetic resonance of so-called plasmonic NPs is particularly suitable for such a fluorescence enhancement effect (Fig. 3.7).

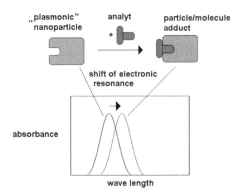

Figure 3.7 Optical sensing caused by shifting resonances of plasmonic nanoparticles by binding of target molecules.

These NPs absorb photons of matching wavelength very efficiently. Noble metal NPs of gold, silver, or composite silver/gold NPs are particularly suitable due to their resonances in the visible range of the electromagnetic spectrum. Silver spheres have a typical resonance around 400 nm. This wavelength can be tuned by variation in the shape and size of silver nanoparticles (SNPs), by mixing gold and silver or by application of composite metal NPs.

High oscillator strengths are observed for the in-axis or the in-plane resonances of nonspherical metal NPs. Thus, gold and silver rods and flat disks and flat triangular nanoprisms are particularly interesting for NP-enhanced fluorescence.

A special demand is the immobilization in such a way that the energy transfer finds optimal conditions. Therefore, the coupling of the fluorophore should be realized at the poles or the corners of the particles. For selective recognition, the binding groups have to be immobilized on these exposed parts of the NP surface, preferentially.

3.2.6 Analyte Detection by Particle Aggregation–Related Fluorescence Quenching

The fluorescence quantum yield of molecules and NPs is not only dependent on the chemical composition of molecules and particles and the molecular structure but also dependent on the environment, on mobility, and on the interaction with neighborhood particles and molecules, too. Among other effects, this is the reason for the frequently observed lowering of fluorescence intensities with increasing concentrations and with the formation of dimers and multimers. In particular, the interaction of fluorophores with metal NPs and the aggregation of fluorescent inorganic NP can cause a drastic reduction of fluorescence quantum yield. In the case of so-called plasmonic NPs, an analyte-induced aggregation of NPs can already be detected by the change of color of the colloidal NP solution (Fig. 3.8).

This effect can be used for the detection of analyte molecules that promote the aggregation of fluorescent particles or act as specific linkers between them. Besides organic fluorescent molecules, even inorganic NPs as fluorescent silicon NPs (Fig. 3.9) can be used for this type of sensing. This principle is the basis for the measurement of heparin by fluorescent silica NPs (Mahtab et al., 2011) dispersed in an aqueous environment (Fig. 3.10). It was found that 3-aminopropyltriethoxysilane (APTES)-modified silicon NPs aggregate by the interaction of heparin with amino groups at the NP surface, resulting in the formation of aggregates and a lowering of the fluorescence signal (Ma et al., 2015).

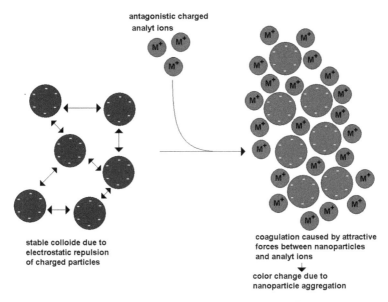

Figure 3.8 Change of optical properties by aggregation of nanoparticles in colloidal solution.

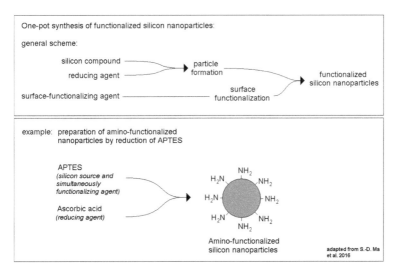

Figure 3.9 Synthesis scheme of amino-functionalized silicon nanoparticles (Ma et al., 2016).

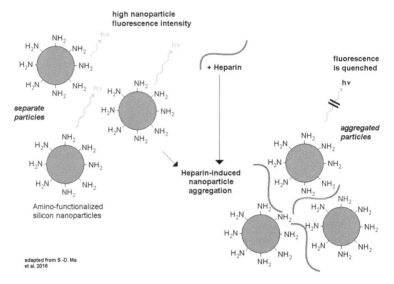

Figure 3.10 Application of amino-functionalized silicon nanoparticles for proof of heparin by aggregation-induced fluorescence quenching (Ma et al. 2016).

3.2.7 Particle-Based Thermal and Optical Sensing of Enzymatic Activities

A general biochemical strategy for bead-based sensing of enzyme activity is the trapping of enzymes at the surface of particles. Proteins can be trapped by surface-attached antibodies. Therefore, the corresponding antibodies have to be immobilized at the particle surface. The enzyme–antibody interaction should take place far enough from the substrate-binding region in order to avoid a disturbance of enzymatic process. The enzymatic activity can be read out by the conversion of a substrate, by the increasing concentration of a product, or by the reaction itself. A convenient non-optical measurement of enzyme activity and kinetics succeeds by microcalorimetric measurements in the case of a sufficient high exothermic character of the reaction.

Colorimetric detection with beads can be performed by the immobilization of polydiacethylene (PDA) on a non-absorbing particle material. Film deposits of PDA react on many changes in the environment by changing their color. This effect can be used for detection of any changes in the composition of a medium with no selectivity. But it can also be applied for a more specific detection if the color change is caused by a specific process, for example, the enzymatic cleavage of a substrate that was immobilized on the particle surface. In particular the activity of a phospholipase was detected colorimetrically by immobilization of PDA on silica microparticles by use of a glycerol-phosphocholine (Nie et al., 2006).

Particles can also be applied for the fluorescence measurement of enzymatic activity in a liquid environment. A simple fluorometric readout can be applied if an enzymatic reaction at the particle surface causes a change in the fluorescence intensity of the particle itself. This effect can be achieved, for example, by an enzymatic cleavage reaction of quencher molecules from the surface of small fluorescing particles. Chemburu et al. (2008) used this principle for the detection of phospholipase activity. They immobilized fluorescence-quenching phospholipids on the surface of fluorescent submicron polymer particles (diameter ~0.5 μm). These conjugates remained stable in the absence of phospholipase. This results in low fluorescence of the particles due to the quenching by the molecules on the particle surface. In the presence of the enzyme, the ester bonds are cleaved and the quenchers are released from the particles surface, which results in an increase in the fluorescence intensity.

3.2.8 Bead-Related Fluorescence Switching

A special interesting strategy is analyte-dependent switching of fluorescence in dependence of the interaction with functionalized metal NPs (Bunz and Rotello, 2010). This method uses the effect that the electronic excitation state of fluorophores is efficiently quenched by a fast energy transfer to gold nanoparticles (GNPs). This technique can not only be applied for detection of smaller molecules but is particularly suitable for the specific recognition of large biomolecules by antibody-functionalized NPs. The strategy can also be transferred to the specific recognition of complete cells and allows to distinguish cells from different organisms as well as the differentiation between healthy and cancer cells (Bajaj et al., 2009). All fluorescence switch-on mechanisms are using an analyte-specific cleavage between a fluorophore (exciton donor) and a quencher (exciton acceptor). Both these components are spatially separated by the analyte-related cleavage, and this process is interrupting the energy transfer between the absorbing fluorophore and the quencher (Fig. 3.11).

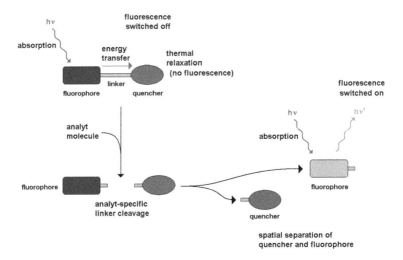

Figure 3.11 Principle of fluorescence switch-on mechanism by analyte-specific cleavage of the linker between fluorescence dye and quencher.

Switch-on strategies are also applicable for the detection of enzymatic activities. This principle was realized for measuring phospholipase activity by the use of quencher-coupled phospholipids immobilized on fluorescent NPs: In the absence of the enzyme, a characteristic emission band of the carrying NP is quenched by an energy transfer. In the presence of the enzyme the phospholipid is cleavage and the fluorescence is enhanced (Cen et al., 2014).

A switch-on strategy was developed, too, by Ang et al. (2014) for the measurement of thiols in a biological environment. They use a thiol-responsive molecule containing a chromophore and a quencher group. These molecules are embedded in a polymer NP. In the absence of thiols the fluorescence is quenched. Thiols switch on the fluorescence under reductive conditions by splitting the disulfide bridge between the fluorophore and the quencher in the sensing molecule. In result, the fluorescence intensity increases with increasing thiol concentration. A similar concept was applied for measuring of thiols by using GNPs as quenchers. In this case, the fluorophore was immobilized at the surface of the metal nanparticles and was realized by the reaction with thiols. This resulted in an up to 40-fold switch-on effect, if the analyte was present (J. Xu et al., 2016).

A conjugation of GNPs with a silica NP and a fluorescence dye by a hairpin-forming DNA oligonucleotide could also be used for the realization of a particle-based assay with a switch-on mechanism (Emrani et al., 2016). Therefore, the fluorescence-labeled hairpin DNA was immobilized on a silica particle, resulting in a fluorescent silica particle. In the presence of a GNP attached to a complementary single-stranded DNA, the hairpin structure on the silica particle is opened, the DNA is hybridized, and the fluorescence is quenched by the metal NP. This adduct can be destroyed if the DNA of the GNP binds stronger to an analyte molecule. This can be achieved if the gold-coupled single-stranded DNA can form an aptamer structure with specific high affinity to the analyte (Fig. 3.12).

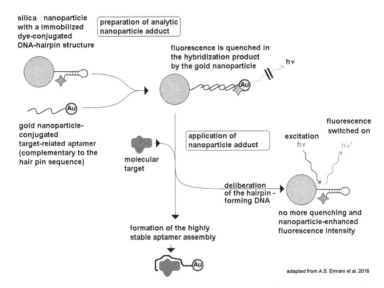

Figure 3.12 Highly specific aptamer-based sensing by a fluorescence switch-on mechanism using a fluorescence-labeled DNA-coupled silica/gold nanoparticle adduct (Emrani et al., 2016).

A coordinative interaction of immobilized ligands with metal ions is a simple possibility for fluorescence switch-off. It can be achieved by the functionalization of NPs by chelate ligands, for example, which are able to bind heavy metal ions (Lee et al., 2014). The principle of fluorescence switch-off by metal ions is illustrated in Fig. 3.13. Vice versa, the removal of the metal ions can be indicated by the switch-on of the particle fluorescence.

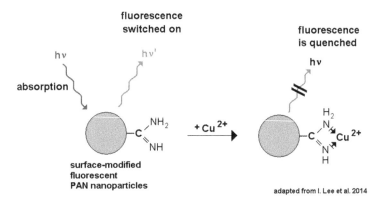

Figure 3.13 Principle of fluorescence switch-off of surface-functionalized fluorescent nanoparticles by metal ions (Lee et al., 2014).

Fluorescence switch-on sensing is also suitable for detection of larger biomolecules if they show certain usable biomolecular activities or can be coupled with them: A connection of an aptamer-labeled graphene quantum dot (GQD) with a molybdeniumdisulfide nanosheet led to a nonfluorescent adduct. The fluorescence quenching is due to the energy transfer effect from the GQDs to the the MoS$_2$ support. The interaction with the analyte molecule causes a release of the aptamer from the MoS$_2$ nanosheet and results in a switch-on effect of the QDs (Shi et al., 2017).

The combination of the fluorescence switch-on strategy with highly specific biomolecular recognition leads to a very powerful method for sensitive detection of biomolecules. This can be used for the proof and for the characterization of cells, too. Duan et al. (2015) reported about QD-based sensor particles. These particles show high fluorescence if they are selectively attached to a bacteria surface, but are quenched if they are trapped by carbon NPs if corresponding bacteria are absent. The authors had been able to show that the bacteria compete successfully with the carbon NPs for binding of the fluorescent sensor particles in the case of a complementary binding factor on the sensor particle surface. The conjugation with highly specific binding aptamers causes a reliable binding between the NPs and specific cell types (Fig. 3.14).

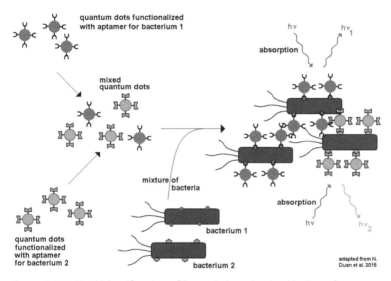

Figure 3.14 Dual identification of bacteria by selective binding of aptamer-functionalized quantum dots (Duan et al. 2015).

Further, it was demonstrated that the application of different quantum dots, the application of two different complemental aptamers and the readout at two different wavelengths allow us to realize a simultaneous dual detection of different bacteria, whereby both the investigated species are surely distinguished. The approach of competing reactions between bacteria and quenching objects with fluorescent particles was also used by Wang et al. (2014) for the detection of different bacteria. Amino-functionalized magnetic NPs have been prepared for the construction of the NP assemblies. These assemblies are disassembled by the bacteria. The authors succeeded in distinguishing groups of several bacteria by the application of up to three different types of sensor particles.

A certain problem of the switch-off technology consists in the fact that fluorescence is highest in the absence of the analyte. This means that small fluorescence differences with a bright fluorescence background have to be detected at low concentrations. This difficulty can be overcome by the switch-on principle.

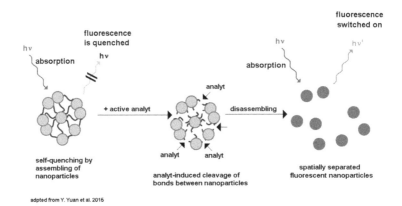

adpted from Y. Yuan et al. 2015

Figure 3.15 Particle-based sensing using a fluorescence switch-on principle by analyte-induced disassembling of self-quenching components (Yuan et al., 2015).

A special way for realizing switch-on effects is the disassembling of self-quenching fluorophore NPs that are spontaneously forming NPs (Fig. 3.15). The disassembling of self-quenching fluorophore NPs by the protein furin is a typical example for the switch-on of fluorescence by a larger analyte biomolecule (Yuan et al., 2015). This furin assay demands for a special furin-activable dye. This dye forms NPs in which the fluorescence is suppressed by particle-internal relaxation processes. Furin initiates the disassembling of these particles resulting in a switch-on of the fluorescence. This method is applicable in living cells and was successfully used for high-resolved imaging of furin activity in MDA-MB-468 cells and for imaging tumor-induced furin overexpression in mice. In a similar way, human serum albumin (HSA) can be detected and used for imaging serum albumin in living cells (X. Fan et al., 2016). It was found that HAS induced the disassembling of NPs formed by an adduct of squareine and pyrene, which can be detected in the ultraviolet-visible (UV-Vis) spectrum as well as by an increase in the NIR fluorescence intensity.

3.2.9 Particle-Coupled Whole-Cell Sensing

A particle-based sensing strategy can also be realized by application of genetically modified microorganisms for the primary signal transduction. The microparticle is only the carrier for the analyte-sensitive cells, in this case. Thus, hundreds or thousands of nonpathogenic arsenite-sensitive *E. coli* cells have been embedded in agarose particles of about 50 µm diameter. Their physiological activity was evaluated by measuring the metabolism-related fluorescence in dependence of the arsenite content of the surrounding medium (Buffi et al., 2011).

The well-established techniques of antibody coupling can be used for realizing specific detection of certain biomolecules by particle-based techniques, too. But, it has to remark that this strategy is not a true sensing method, because the antigen–antibody interactions are not reversible in the meaning of a molecular sensor. The required molecules for antibody couplings are large proteins. There low ability for diffusion through the matrix is the reason why they are normally not incorporated inside a particle matrix but preferred immobilized at the particle surface. The surface immobilization ensures the free access of analyte molecules as well as of fluorophores and complementary antibodies.

3.2.10 FRET Transduction

FRET technology is also directly related to the electronic excitation by resonant interaction of chromophores with electromagnetic waves. The signal is also given by the emission of photons of higher wavelength as in the case of fluorescence and phosphorescence. But, there is an additional step in the conversion of energy.

Normally, electronically excited molecules have three main paths for the spontaneous deactivation. The most frequent path for pigment dyes is the vibrational relaxation down to the electronic ground state. The absorbed photonic energy is converted completely into mechanical energy, and therefore heat, by this mechanism. The second relaxation way is represented by the radiating deactivation— either directly via fluorescence or via a triplet state and deactivation by releasing a photon of phosphorescence light. In most cases of radiating deactivation, a certain amount of the electronic excitation energy is released by vibrational relaxation, too. This amount of energy corresponds to the difference between the energies of the absorbed and the released photon. A third relaxation way can run over a chemical reaction. In this case, a part of the excitation energy is required for the reaction activation and contributes to the change of internal energies of reactants and products of the photochemical reaction. The difference between the electronic excitation energy and this change of internal energies during the photochemical reaction is also dissipated in the form of heat.

In the FRET process, the excitation energy is transferred, at first to another chromophore. This process is normally accompanied by a low energy loss due to a small vibrational relaxation in the electronically excited state. The energy transfer from the primary absorbing molecule to the FRET-receiver is a resonant process. This means that energy difference between the electronically excited state and the ground state of the absorbing molecules is equal to the energy difference of the electronically excited state and the ground state of the receiver molecule. After this energy transfer this secondary excited molecule is able to relax electronically by emitting a photon or—in the case of simple electronic quenching—by vibrational relaxation to the ground state. The radiative decay (emission) corresponds to the release of a photon in the case of fluorescence.

The transfer of excitation energy is dependent on the spatial distance between the sending and the receiving chromophore. Therefore, the luminescence intensity in the FRET process is distance sensitive. Thus, FRET can be used for determining the mean distances between both the chromophores quantitatively. This possibility is very interesting for detecting molecular binding events and for getting information of conformational changes.

FRET is applicable for detecting of binding events on the surface of microbeads, too. Specific binding events on particles can be detected by driving out FRET-active molecules from a surface during the substitution by a specific analyte, for example. Thus, FRET can be used for the imaging of reductive environments of living cells, whereby the required high emission intensities can be achieved, for example, by the application of dye-doped silica NPs (Petrizza et al., 2016). Meanwhile, a large spectrum of sensor NPs with signal transduction by FRET is reported (Shi et al., 2015), among them being compound semiconductor QD NPs, GQDs, graphene oxide nanoparticles (GO NPs), and GNPs.

Naphthalimidazide-modified carbon NPs have been developed for the fluorimetric detection of H_2S by a FRET process (Yu et al., 2013). In this case, an emission at 526 nm is observed only after the reduction of the azide groups by H_2S due to an energy transfer from the electronically excited carbon particles to the naphthole chromophore.

FRET-functionalized silica NPs were used for measuring the activity of metalloproteinase II (Achatz et al., 2009).

SNPs have been applied for the enforcement of FRET signals reporting about posttranslational modification of proteins a sialic acid terminal oligosaccharide. Therefore, the acceptor fluorescence dye (Cy5) was connected with the target carbohydrate by a multistep transduction chain. The donor dye was coupled with the SNP. This complex was specifically connected with the target glycoprotein molecule by specific aptamer–protein recognition (Fig. 3.16). The authors argue that an enhancement of the FRET-based fluorescence is due to the plasmonic properties of the SNPs (T.B. Zhao et al., 2017). A similar strategy was applied for whole-cell recognition (pathogenic bacteria) via an aptamer–surface protein reaction in the presence of GNPs (Jin et al. 2017).

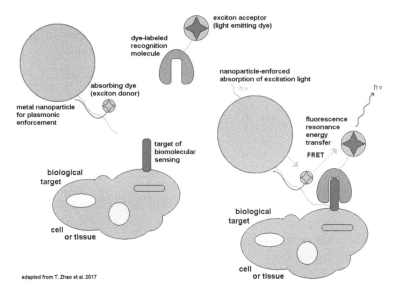

Figure 3.16 Particle-based FRET system for specific detection of biomolecules on cell surface for identification of biological cells (T.B. Zhao et al., 2017).

A general approach for protein-related chemical or biochemical sensing was realized by coupling of semiconductor NPs (QDs) with fluorescent proteins. The QDs are used as energy donors and the fluorophores of the proteins as energy acceptors in a FRET process. Boenemann et al. (2012) conjugated CdSe/ZnS core/shell NPs with the fluorescent proteins YFP, mCherry, and b-phycoerythrin, a light-harvesting protein complex.

Besides the transduction of molecular signals, FRET is also very useful for the spatial-resolved readout of local temperature. Such a transduction can be realized by a resonant energy transfer between compound semiconductor NPs (quantum dots) and organic dyes. Albers et al. (2012) showed a temperature resolution of about 0.2 K by a FRET process between hybrid quantum dots of CdSe nanodots and CdS nanorods as energy donors and a cyanine dye as acceptor. They used these nanothermometers for in situ measurement of temperature differences on HeLa cells.

The application in barcoding and labeling is an additional application of FRET NPs. This application uses the enlargement of Stokes shift by the exciton transfer. This is a specific advantage of the FRET mechanism and is due to a certain reduction of excitation energy in the whole energy transfer process. This energy shift is caused by the orthogonal Franck–Condon transition in the electronic excitation and the subsequent fast oscillation relaxation. This process can going on two times in the case of the FRET mechanism—one time in the excitation of the donor dye and a second time in the energy transfer to the acceptor dye. In result, exceptional large distances (Stokes shifts) between the absorption maximum (donor dye) and the emission maximum (acceptor dye) take place that allow an easy spectral separation of emission light from the excitation light. Further, it is easier possible to design systems with the same excitation wavelength but a different Stokes shift by combining one donor dye with different acceptor dyes. This allows one to construct NPs with different emission colors that are excitable at the same excitation channel and to mix different emission colors in one NP. Thus, sets of different multicolor NPs can be generated by a combination of acceptor dyes. Silica NPs are particularly suitable for this purpose. The number of distinguishable NPs can further be enhanced by different concentration ratios of two or more donor/acceptor pairs. Wang and Tan (2006) realized multicolor FRET microparticles of a diameter of about 6 μm with hundreds of surface-attached NPs of 70 nm with up to five different emission colors.

3.3   Chemoluminescence Transduction

Chemoluminescence is the effect of nonthermal generation of photons by a chemical reaction. This type of reaction is based on the phenomenon of forming a chemical product from a thermally activated reaction—that could be also a reaction at room temperature, for example—in the electronically excited state. Such a formation of electronically excited products is possible from an energetic point of view due to the fact that many exothermal chemical reactions are releasing energies per molecule in the same order of magnitude like the energy of visible photons (Fig. 3.17). The whole process corresponds completely to the quantum mechanical principle of one-molecule-one-photon resonances.

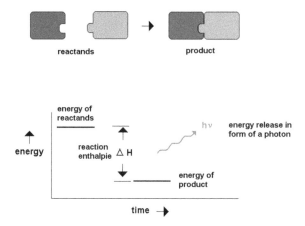

Figure 3.17   Principle of chemoluminescence: Due to the exothermic reaction, the product of a chemical reaction is formed in the electronically excited state, and this energy is dissipated in the form of an emitted photon.

In contrast to fluorescence, phosphorescence, and FRET sensing, chemoluminescence does not require any exciting light. The energy source is a chemical or biochemical dark reaction. This has the big advantage that measurement of light intensities from chemoluminescence can proceed under absolute dark conditions, and a complete exclusion of parasitic or background light is ensured. This effect makes chemoluminescence particularly suitable for highly sensitive measurements.

Chemoluminescence reactions are often catalytical reactions. In living nature, such reactions are catalyzed by enzymes. The high catalytical efficiency of chemoluminescence enzymes makes it possible to generate light energy originating from biochemical processes in cells at room temperature. The high substrate specificity of enzymes makes these processes highly selective and, therefore, well applicable for biochemical sensing.

For detection of a certain substrate by particle chemoluminescence sensing, the enzyme is immobilized at the surface of the particle. If the substrate and each required cosubstrate are mobile enough and the particle matrix can sufficiently swell, then the enzyme can also be incorporated inside the particle matrix. In the presence of the matching substrate, the enzyme-loaded particle becomes bright due to the ongoing light-generating reaction. Konno et al. (2006) reported polymer NPs with a phospholipid layer and immobilized luciferase for generating chemoluminescence from the NPs in the presence of adenosine triphosphate (ATP), luciferine, and oxygen.

3.4 Transduction by SERS

Molecular sensing by visible or UV light in condensed phases calls mostly for labeling techniques. A label-free UV-Vis-spectrophotometry or even fluorescence supplies less specific signals because of the comparatively broad optical absorption and fluorescence bands, which are typical for aqueous solutions and other liquids. Much smaller absorption bands are supplied by vibrational spectroscopy. This technique is suitable for the identification of specific chemical groups, and patterns in the spectra can be used for the identification of substances or mixtures by fingerprint signals. The electromagnetic resonances of these molecular motions are in the middle-infrared (MIR) range, and therefore, IR spectroscopy is the standard method for characterization of substances by their vibration characteristics. Unfortunately, the application of IR spectroscopy for biological objects is limited because of the low IR transparency of aqueous solutions. An interesting alternative is given by Raman scattering. This method allows the application of visible light for recording of vibrational spectra. The Raman effect is caused by a certain transfer of energy between the optical photons and the vibrational energy of molecules. It is in the nature of Raman scattering that the signals are low and, normally, only substances at higher concentrations can be measured by Raman spectroscopy.

This restriction was overcome when the surface-enhanced Raman effect was discovered. It was found that the interaction of molecules with metal surfaces, in particular with silver and gold, results in strong enhancement of the Raman signal. The enforcement factor of surface-enhanced Raman spectroscopy (SERS) is several orders of magnitude, which means that substances in the micromolar or nanomolar range and sometimes at still lower concentrations can be observed.

The intensity of SERS signals depends on the surface area. The specific surface is much higher for distributed small particles than for a smooth metal surface. That's why gold and silver NPs are very useful for SERS measurement. They can be applied for chemical characterization, but they can also support the characterization of cell cytoplasma if they are introduced into cells for acting as SERS nanospies. The fingerprint character of Raman spectra is very specific and allows to distinguishing bacterial species, for example (Petry et al., 2003).

Gold and silver NPs are well suited for SERS sensing. They can be prepared easily by standard chemical methods and can be introduced into biological material as well as into microfluidic systems. Thus, GNPs have been used for SERS readout of molecular information from microfluid segments (Strehle et al. ,2007). Microfluidics and in particular microfluid segments are particularly suitable for the use of dispersed noble metal NPs for SERS sensing (März et al., 2011). The Raman spectra from single microfluidic compartments (Fig. 3.18) can report on changing concentrations or changing composition of microfluid segments in complex experiments, for example, in chemical syntheses or in microbiological cultivation experiments.

Silver surfaces are still better suited for Raman enforcement than gold surfaces, but their application suffers from the oxidation sensitivity of silver. A compromise is the deposition of a thin gold layer on the surface of silver nanocores. Such gold/silver core/ shell particles had been successfully applied for SERS detection of aromatic compounds as naphthothiole (Samal et al., 2013).

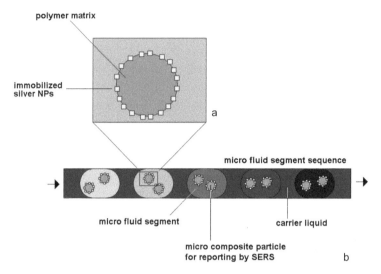

Figure 3.18 Application of SNP-containing SERS microsensor particles for characterization of chemical composition of sequences of microfluid segments.

The analytic information that can be obtained from NP-based SERS sensing can be very valuable and very specific. SERS has the potential to develop to a universally applicable key method for a next-generation technology for monitoring chemical processes in cells, tissues, and living animals (Kneipp, 2017). Besides classical SERS measurements, two-photon excitation promises the development of high-spatial-resolution techniques in the frame of so-called hyper–Raman scattering and surface-enhanced hyper–Raman spectroscopy (SEHRS). High specificity and sensitivity can be connected by use of the combination of specific molecular interaction with spectral properties of NPs and molecules. A special strategy is the application of dye molecules that bind on analyte molecules at the surface of dispersed SNPs. This technique could be applied for the sensitive recognition of antibiotics by a shift of SERS signals due to the molecular interaction (März et al., 2012).

The interaction of cells with noble metal NPs allows also an identification of biological material. Raman spectroscopy was identified as a very efficient tool for probing a large spectrum of organic material from all organisms and is even applicable for the characterization of single bacteria (Petry et al., 2003; Pahlow et al., 2015). A discrimination of different bacteria strains succeeded by the application of SNPs for SERS measurements in a microfluidic chip device (Walter et al., 2011). The applicability of SERS for identification of microorganisms in microfluidic devices was also demonstrated by the investigation and differentiation of six mycobacterium species (Mühlig et al., 2016). The combination of specifically binding SERS-active particles with magnetic activity is very attractive for bioanalytics and medical diagnostics. The specific recognition can be realized by particle-immobilized biomolecules. Thus, aptamer-conjugated magnetic beads can capture, for example, disseminated tumor cells, and the binding can be detected by SERS measurements (C.L. Sun et al., 2015).

A certain problem for the application of metal NPs for SERS analytics is their handling. In principle, they can be dispersed in a solvent by forming a colloidal solution. But, the colloidal state is sensitive against chemical changes. Increasing salt concentrations and sometimes surfactants, peptides, or other additives can cause a collapse of the colloid, aggregation of particles, coagulation, and precipitation. In addition, NPs tend to adsorb and can be trapped by interfaces. Finally, the direct interaction with cells is nondesired in many cases because of toxic effects.

These problems can be overcome by constructing composed polymer microparticles for SERS sensing (Köhler et al., 2013). The SERS-active NPs are either deposited at the surface of a polymer particle (type I) or integrated inside the swellable or gel-like polymer matrix (type II). The immobilized metal particles remain their ability to interact with molecular components of a liquid environment and are well suited, therefore, for SERS analytics (Visaveliya et al., 2015a). But the NPs are fixed and stabilized, cannot aggregate with each other, cannot adsorb on other surfaces, cannot be taken up by cells, but can be manipulated together with the carrying microparticle. Mostly silver and gold are applied in polymer composite NPs for SERS applications. In addition, alternative composite microparticles have been obtained by the deposition of ZnO on poly(amic acid) and proposed for analytical purposes (Wu and Hsu, 2015).

The surface of a microparticle with a diameter in the upper-micrometer or submillimeter range can carry several millions of metal NPs. Several billions of the small metal NPs can be incorporated in the volume phase of such a particle. Thus, a single microparticle can supply a strong SERS signal and is usable as a single-particle sensor. Such polymer composite particle sensors can easily be placed in microchannels or capillary tubes in order to realize a microflow-through SERS analysis arrangement (Fig. 3.19).

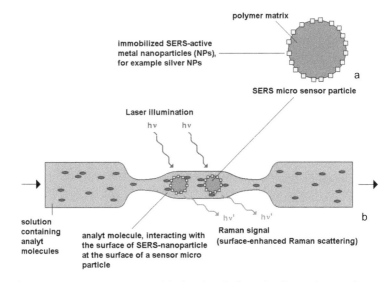

Figure 3.19 SERS sensor particle for chemical sensing in a microcontinuous flow process.

A general problem in SERS sensing is the dilemma between the desired adsorption and accumulation of analyte molecules on the metal surface, on the one hand, and the desired reversibility of analyte binding for a continuous use of the SERS effect of the metal surface, on the other hand. In the case of high binding constants (strong binding), a high accumulation effect and, therefore, high sensitivity, but no reversibility, are achieved. In the case of low binding constants, the reversibility is high due to the concentration-dependent adsorption and desorption of analyte molecules, but the accumulation effect and the sensitivity are low.

This dilemma can be overcome for some types of analyte molecules by a periodical chemical regeneration of the surface of the sensor particle. This strategy is applicable, for example, for amines and other nitrogen-containing organic molecules, which are frequent analytes in bioanalytics. They have a high affinity to silver surfaces and show high accumulation effects but can be released from the surface by protonation. Thus, a simple acid cleaning is sufficient for the regeneration of the particle surface. The principle of alternating measurement and regeneration phases was successfully implemented in a microfluidic analysis arrangement by a simple fluid switching (Visaveliya et al., 2015b).

SERS sensor particles of type I can be prepared by surface binding of SNPs. The immobilization can be realized, for example, by electrostatic forces. Negatively charged SNPs are fixed at polymer surfaces with positively charged groups of the matrix itself, by surface-attached polycationic macromolecules, or simply due to a previous binding of cationic surfactant molecules on the polymer surface.

Type II SERS sensor particles have to be prepared by a polymerization step in which the NPs are introduced. This can be achieved by mixing a colloidal solution of metal NPs with the liquid monomer or by redispersing of metal NPs directly into a liquid monomer mixture. This strategy is suitable for generation of water-swellable gel-like particles, where the monomers can be polymerized and cross-linked in an aqueous solution containing the metal NPs in a colloidally dispersed form.

A monomer mixture containing acrylamide with an addition of bisacrylamide in aqueous solution is well suited for the generation of the type II particles. Sensor particles with a diameter in the submillimeter range and homogeneous size can be prepared by microfluidic droplet generation and photochemical flow-through initiation of polymerization. A combination of surface attaching of SNPs and higher loading can be achieved by hierarchically constructed three levels (Visaveliya et al., 2017). For preparation of this particle type, first, SNPs are immobilized on the surface of submicron polymer particles. These particles can be formed by no or less swellable material, for example, polymethylmethacrylate (PMMA). Second, these particles are deposited like a shell on the surface of a submillimeter polymer particle. The film of silver-loaded submicron particles combines a local high concentration of SNPs with good access of analyte molecules from a solution through the pores of the particle film to the silver surface.

Inorganic composite particles containing noble metal NPs for SERS sensing can be prepared by a nanocapsule technique: In this case, an inorganic shell like silica was deposited at the surface of the polymer particle. In the second step, GNPs were immobilized at this inorganic shell, and finally the whole particle was calcinated in order to remove the organic material. In result, an inorganic capsule containing SERS-active noble metal NPs was obtained. Then, the metal NPs can be functionalized for special sensing applications, for example, by p-aminobenzenethiole for the detection of nitric oxide (Rivera-Gil et al., 2013). The deposition of a chitosan layer on silver nanotriangles led to SERS sensor particles suitable for sensitive in situ SERS measurements and for SERS imaging of cells (Potara et al., 2013). This technique allows us to obtain a series of microscopic images of a cell or tissue at different Raman wavelengths. The high sensitivity and spectral resolution of NP-supported SERS measurements supply detailed information about the spatial distribution of substances causing different Raman peaks.

3.5 Plasmonic Particles for Signal Transduction

Metal NPs show characteristic electromagnetic resonances. These resonances are due to the electronic excitation of mobile electrons. These excitations are called particle plasmons in analogy to localized electronic excitation states at metal surfaces and thin metal films, which are also characterized by specific resonances. Usually, metal NPs with such characteristic absorption peaks are called plasmonic particles. The energy of these plasmonic resonances is dependent on the element, on the size, and on the shape of the particles and on the properties of their environment (Kneipp et al., 2015). Despite the fact that the application of the model of plasmons might be less relevant for understanding the physical nature of the electromagnetic particle resonances in comparison with the assumption of a molecule-analog one-photon-one-electron process, the term "plasmonics" is used for such sensor particles hereafter with respect of its importance and standard-like use in the literature.

Investigations on the application of metal NPs in assays for biodiagnostics had intensified during the 1990s. In particular, the conjugation of metal NPs with DNA and its use in specific DNA recognition by hybridization was studied (Fritzsche and Taton, 2003). Silver and gold NPs are particularly in the focus of particle plasmonics because their resonances are in the visible part of the optical spectrum. Particular high-resonance wavelengths are observed in the case of nonspherical gold and silver NPs, for example, rod-like, disk-like, or flat prismatic particles. For high aspect ratios, the peak maxima can be shifted up to the NIR region.

Silver nanospheres have a plasmon resonance around 400 nm. Thus the combination of silver and gold in binary metal NPs in combination with a nonspherical shape allows the tuning of the resonance energies over the complete visible range by variation of the gold:silver ratio, the shell thickness of core/shell particles, and the aspect ratio of nonspherical binary particles.

Besides the metal composition and the particle shape, the environment of the plasmonic particle has an effect on the resonance energy, too. In particular, the refractive index of the environment influences the position of the absorption maximum. In the case of small plasmonic particles, even a few surface-attached biomacromolecules can cause a significant shift of the resonance wavelength. In addition, electrical charges—for example, by ionic molecules adsorbing at the particle surface—cause a shift in the resonance energies. This shift can be suitable for sensing. In principle plasmonic measurements on specifically functionalized noble metal NPs can be used for getting detailed information, for example, by stereo- and chiroselective resonances or by molecular imprinting techniques (Riskin et al., 2010). It is applicable, for example, for proving antibiotics by use of NPs (Frasconi et al., 2010). But, it has to be kept in mind that the spectral and intensity effects are comparatively low and demand for well-controlled conditions and high precision in the optical measurements.

In the simplest case, plasmonic NPs are used for detection of a change in refractive index of a solution. This information is less specific, and a comparatively high shift is required for generating a significant wavelength shift. Much more interesting is the detection of specific binding of molecules. This can be achieved by coupling the plasmonic NPs with recognition molecules like oligonucleotides for the binding or DNA with complementary base sequences or antibodies for binding of specific antigens. The following transduction chain can be constructed:

Analyte molecule → Binding on the NP surface → Shift of resonance wavelength → Noncontact optical detection by measuring the change in absorbance at a fixed wavelength (monochromatic photometry) or measuring the change of absorbance on two fixed wavelengths and calculation of ratios of absorbance differences or measuring the shift of the resonance maximum (spectrophotometry)

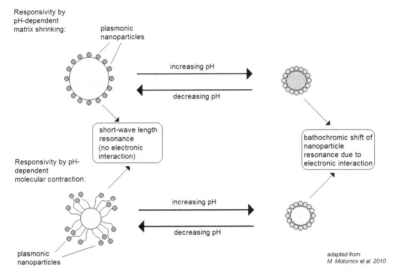

Figure 3.20 Principle of pH measurement by variation of electronic coupling between plasmonic NPs immobilized on the surface of pH-dependent swellable and shrinkable polymer particles (Motornov et al., 2010).

The resonance wavelength of plasmonic NPs is dependent from their size and is also shifted if smaller particles come into close contact, which allows an electronic interaction between neighborhood single metal NPs. By this effect, the shrinking or swelling of polymer or gel-like microparticles causes a spectral shift of plasmonic resonance of metal NPs that are incorporated inside the polymer matrix or on its surface (Fig. 3.20; Motornov et al., 2010). Similarly, the contraction or expansion of linker molecules between a microparticle and plasmonic NPs can be applied for sensing. Both effects are used for the realization of pH-sensitive environmental-responsive polymer particles (Lupitskyy et al., 2008).

In contrast to SERS sensing, the measuring of transmitted or remitted light occurs at the illumination wavelength. Complete spectra for detecting the shifts of maxima can be recorded either by a tuning of the irradiation wavelength or by the application of polychromatic light, followed by dispersion and recording of transmitted or remitted intensities for all wavelengths, for example, by a fiber-coupled compact spectrometer.

Besides NPs of a single material, composed spherical and shape-anisotropic metal NPs can be applied for biomolecular sensing. Silver core/gold shell particles are attractive for plasmonic sensing due to the sensitivity of their plasmon resonance against environmental conditions. This kind of particle sensorics is based on the so-called localized surface plasmon resonance (LSPR) and is promising for detection of changes in refractive index (Steinbrück et al., 2011).

The application of the required plasmonic NPs for sensing is analogous to the SERS sensor particles. The metal particles act as primary transducer, and therefore, they have to come in close contact with the analyte. Thus, they can be applied as dispersed particles in colloidal solution or fixed at wall surfaces or at surfaces of carrier particles, or they can be incorporated inside a gel matrix, which is permeable for the diffusing analyte molecules. An enhanced selectivity for the interaction with analyte molecules can be achieved by the embedding of the metal NPs in a molecular-imprinted polymer (MIP) matrix (Fig. 3.21; Ahmad et al., 2015). Selective sensitive metal/polymer composite particles have been realized by embedding the plasmonic particles in imprinted polymer microparticles (Wu et al., 2012). The application of GNPs for measuring the pH inside mammalian cells succeeded by encapsulation with a robust poly(vinyl alcohol)-polyacetal shell (Stanca et al., 2010).

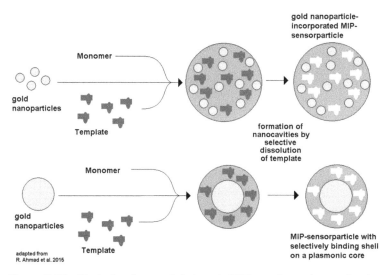

Figure 3.21 Strategies for particle-based SERS sensing using molecular-imprinted polymers incorporating GNPs: co-embedding of template molecules and GNPs in the polymer matrix (top) and molecular imprinting of a thin polymer shell on a gold core (bottom) (Ahmad et al. 2015).

He et al. (2013) found that the activity of an enzyme can be detected by a shift of the plasmon resonance peak if the enzyme molecules are deposited at the surface of the noble metal NPs. For this purpose, they synthesized Ag/Au core/shell particles and immobilized glucose oxidase (GOD) on these particles. An LBL-analogous technology was applied for the protein immobilization. The GOD was linked to the metal surface by a layer of poly-L-histidine (PLH). Each added layer on the particle surface caused a bathochromic shift of the plasmon resonance. Whereas the addition of PLH is reflected by a slight shift only, the addition of GOD caused a moderate shift, and a further strong bathochromic shift was recognized by the addition and binding of the substrate glucose. The shift of the resonance band was large enough, could be measured reproducibly, and gave the authors the possibility to measure the kinetics of enzymatic reactions.

A big drawback of plasmonic sensing is the low effect of environmental refractive index changes on the resonance wavelength. In general, the caused shift amounts to only a few nanometers or even part of a nanometer. Therefore, the synthesis and application of nonspherical and composite NPs is under investigation in order to find configurations with an enhanced effect on the absorption maximum and high sensitivity, even at a low number of molecules binding to the particle surface.

3.6 Particle-Mediated Signal Enforcement at Interfaces

Optical detection methods that are used in analytical procedures or sensing are mostly mass sensitive. Molecule-number- and therefore mass-related effects are not only important for absorption and fluorescence but, for example, also important for the effects of reflectivity, light scattering, refractive index changes on interfaces, wave guiding of thin films, and wavelength-related filter functions. The intensity of optical signals from the attachment of molecules can be enforced if the size or mass of bounded objects is enhanced. Significant enhancement can be achieved if selectively binding NPs can be used for this enforcement. The effect of a single particle on the signal enhancement is strongly dependent on the particle properties and the detection method.

In contrast to the aforementioned optical methods, in which the particle acts as a converter of optical signals, particle-mediated signal enforcement means that a transduction effect is already present without particles but enforced by them. Typically, the particle has the character of a label that is attached to an analyte molecule that is attached to an interface. The sensing chain has to include several coupling components: At first, the analyte molecule has to bind selectively to the surface. The specificity can be achieved by immobilization of suitable molecular recognition elements as antibodies for the detection of proteins or oligonucleotides for hybridization of DNA or RNA containing specific sequences, for example. In a second step, a labeling procedure has to take part on all sites on which a molecular binding occurred. For this labeling procedure, a high specificity is not required, but it should be designed in a way that prevents the binding of the labels on surface parts without bonded analyte molecules. Otherwise, the nonspecifically bonding labels would generate a large background signal and enhance the noise in the sensing process. In the case of analyte molecules of a certain size, the binding of a complementary antibody could be a suitable general strategy. A certain signal enhancement can already be caused by this complementary antibody itself due its considerable molecular mass. But, a higher enforcement can be achieved if the complementary bonding is used for connecting the surface-attached analyte molecule with an NP in order to forming an adduct with the complementary antibody.

This particle enforcement is well applicable for special highly sensitive optical transduction at interfaces, for example, surface plasmon resonance (SPR) and reflective interference spectrometry (RIFS) measurements (Hanel and Gauglitz, 2002). Both techniques are strongly reacting on enhancement of the interface mass loading. In SPR, the signal shift by mass loading is due to a shift in the resonance angle of maximum coupling of a laser beam into a plasmon-conducting surface film. These changes in angle can be read out very sensitively. A small number of bonded particles are sufficient to supply a significant signal, and in some cases, the detection of a single binding event might be detectable.

In RIFS technology, the binding of molecules or NP at an interface shifts the thickness of light passing through an optical interference layer. Such thickness changes can be detected by sensitive measuring small changes in the intensity of transmitted light. By this approach changes in the mean layer thickness of less than 1 nm are detectable. Thus, this method is also applicable of the detection of binding of submonomolecular analyte films or a low number of binding particles. A promising new strategy is the coupling of molecular-imprinted polymer nanoparticles for the detection of specific analytes with RIFS technology, which was successfully applied for sensitive measurement of antibiotics (Weber et al., 2018).

The attachment of larger NPs can also be used in order to enhance light scattering from smooth interfaces. Metal NPs are particularly suitable for generating high-scattering effects.

The problem of this scattering light detection strategy represents a certain contradiction between sensitivity and specificity: Large particles support high sensitivity because they have much stronger scattering than smaller particles, on the one hand. On the other hand, larger particles have a higher probability of nonspecifically bonding to the interface in comparison to smaller particles. This contradiction between sensitivity and specificity is a general problem for particle-mediated signal enforcement.

3.7　Counting Binding Events by Particle-Mediated Nanolight Valves

A particular high sensitivity of particle-mediated sensing can be achieved with special arrangements for light passage through a nontransparent film. The trick is the use of a small opening in such a film by small holes. Holes, which are significantly smaller than the wavelength of incoming photons, have only a low probability of penetration by these photons. The majority of incoming photons are reflected or absorbed by the film and only, from time to time, a single photon can pass through the hole. The interaction of photons with the hole in the film could be understood like a dropping of water through a mechanical hole. The penetration probability decreases strongly with decreasing size of the hole.

The binding of single molecules has nearly no effect on this photon dropping. But the attachment of an NP on the hole can modify the situation significantly (Fig. 3.22). It was observed that bonding particles can enhance the probability of photon penetration through the holes by several orders of magnitude. This gives the possibility to generate high signal-to-noise ratios for single-particle-binding events.

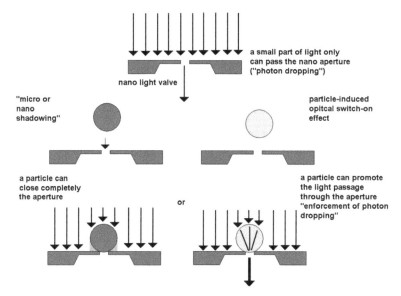

Figure 3.22 Principle and device for nanolight valve sensing using the binding of nanoparticles at the position of a small optical aperture.

In principle, this strategy is suitable for the application of single-molecule detection if a molecular binding event can be coupled with the fixation of a particle in the hole. This can be achieved by the afore-described complementary recognition strategy: At first, the analyte molecule has to be recognized by a suitable receptor that is attached at the hole position. Second, the particle has to be attached after binding of the analyte molecule by an analyte-complementary molecular functionalization of the NP. The binding of the particle at the hole position strongly modifies the light transmission behavior. Either the hole is closed by the particles and the light path is blocked, or the particle promotes the passage of photons, resulting in enhanced brightness of the hole.

3.8 Nanoparticle-Supported Photothermal Effects and Photoacoustic Imaging

In general, NPs can be conjugated with biomolecules and used for highly specific interactions. The high compatibility with biomolecules makes smaller NPs particularly interesting. They can be used for different strategies of improvement of molecular imaging, for example, by the local conversion of energy (Debagge and Jaschke, 2008). Thus, light-absorbing NPs are well suited for local conversion of light energy into heat. This photothermic effect can either be used for actuation or chemical activation as well as for sensing or imaging purposes. The energy conversion by NPs is connected with an energy-focusing effect because the absorbing NPs can be much smaller than the wavelength of incident light. The absorption of radiation by an NP causes the formation of a local hot spot. In consequence, at least, a local lowering of refractive index is caused. In addition reversible as well as nonreversible chemical changes can be induced, which enhance the local contrast and supply information about the imaged object.

NPs of heavy metals are of particular interest for these photothermic effects due to their high efficiency in light absorption. So-called plasmonic NPs as consisting of gold or silver show very high and characteristic absorption caused by their plasmonic resonance. The effect of thermoplasmonics can be used, for example, for particle-based photoacoustic imaging (Baffou and Quidant, 2013). Differences in local chemical properties or biomolecular affinities can be made visible by surface functionalization of these particles, causing accumulation of the contrast-enhancing particles in the regions of interest and visualization of certain targets in cells and tissues.

A special technique of imaging of tissue structure was realized by the introduction of NPs into the tissue and measuring the effects of particle distribution on photoacoustic signals (Yang et al., 2009). The photoacoustic signals are initiated preferably by a laser pulse. The deliberated heat is not only simply dissipated into the environment, but it can initiate a microscaled mechanical shock wave propagating through the object and the substrate. This effect increases with increasing laser power and focusing. Yang et al. (2014) used copper sulfide NPs by injecting them into mice tissue. This technique allows us to get time-resolved tissue images by applying series of laser pulses and photoacoustic measurements. It can be used for the characterization of local enzyme activity in tumor tissue, for example.

Semiconductor compound NPs (QDs) are promising for combining photoacoustic imaging with photothermic effects for contrast enhancement (Shaskov et al., 2008). Such a combination was realized, too, by gold-plated carbon nanotubes (Kim et al., 2009) and by NPs consisting of a gold core and a palladium shell (M. Chen et al., 2014).

Despite inorganic materials, polymer particles (Chen et al., 2017) and other particles of organic substances, among them preferably polymer-encapsulated organic NPs, can be applied in photoacoustic imaging (Li and Liu, 2014). The selection of an optimal spectral channel for the photoacoustic excitation is an important aspect of the application of this method. The near-UV range, the shorter-wavelength visible range, and—in some cases—the longer-wavelength visible range can be critical for the required focused laser illumination. In these cases, an excitation in the NIR region can offer an attractive alternative (Jiang and Pu, 2017). This spectral range can also be of particular interest for the combination of imaging and photothermal heat therapy with self-assembling organic NPs (Li et al., 2017).

NPs of heavy elements, namely noble metal NPs, are favored for different labeling purposes in sensing and bioimaging but are also of interest for therapies based on local heating of tissue. Gold nanorods are used, for example, for localized photothermal stimulation. The effect improves the microscopic imaging of tissue and can be used, at the same time, for cancer treatment by laser-induced local heat delivery, cell disruption, calcium deliberation, and homeostasis (Tong et al., 2009). Jokerst et al. (2012) reported the introduction of gold nanorods into ovarian cancer tissue for photoacoustic imaging. The intensive interaction between laser radiation and absorbing NPs promotes their application for guidance of local heat treatment of tissue by photoacoustic imaging. In these applications, the NPs serve for the therapeutic conversion of light into local heat, on the one hand, and for the contrast enhancement during the microscopic process control and monitoring, on the other hand.

This combined function can advantageously be coupled with tumor-selective functionalization of the NPs (L.H. Du et al., 2017). Gel NPs consisting of polyanilin-loaded gamma-polyglutamic acid had been developed for this purpose and had been reported to be well suited for photothermal and photoacoustic transduction, as well as marked by excellent biocompatibility (Zhou et al., 2017).

Biomedical motivations stand mainly behind the recent investigations of particle-assisted photoacoustic imaging. But besides the human body, other organisms can be interesting objects for photoacoustic investigations. Thus, photoacoustics was used for the characterization of NPs inside different parts of plants, too (Khodakovskaya et al., 2011).

Besides massive metal NPs, hollow NPs and nanocapsules (An and Hyeon, 2009) are applicable for photoacoustics, too.

3.9 Particle-Based Reversible Sensing by Photochemical Switching

High specificity and selective binding demand high binding constants in general. But, high binding constants represent a problem for sensing, because high binding constants contradict the desire for reversibility. Washing or chemical regeneration is often unavoidable in order to reuse a chemical sensor after loading with an analyte.

The application of light for bringing a sensor in a sensitive state is a convincing strategy for miniaturized sensing with high specificities. Scarmagnani et al. (2008) took this principle for the development of a repeated working particle-based sensor for Cu^{2+} ions. Their development was based on the fact that spiropyranes can photochemically actuate for switching between a high-conjugated (open state) and a low-conjugated state (closed ring), on the one hand, and on their state-dependent binding of metal ions, which can be detected by a color shift, on the other hand. Copper is bonded with high affinity from the open molecular structure, resulting in an intensive color, which can easily be detected. This copper can be released from the chelating molecule if it is photochemically switched into the closed ring structure (spiro-state). The application of this principle succeeded by the immobilization of the spiropyrane photoswitches on the surface of polystyrene (PS) beads of ~2 μm diameter.

Besides direct sensing, NPs with photoswitchable fluorescence are interesting tools for on- and off-switching of luminescence (M.-Q. Zhu et al., 2005). This effect could be used for the switching of reference signals in a small liquid volume or in tissues and cells, for example. Such particles had also been synthesized in the size range between 40 and 400 nm by incorporation of spiropyran-merocyanin dyes into the polymer matrix of particles (Zhu et al., 2006). Spiropyran-functionalized monomers have been applied for the direct synthesis of photoswitchable NPs by polymerization (Zhu et al., 2011). These particles had been successfully applied for two-photon photoswitching. After immunofunctionalization, these particles are well suited for the selective imaging of cells in tumor tissues.

3.10   Hierarchical Particle Architectures for Optical Sensor Applications

There are several motives for assembling particles for sensing. The simplest motivation is the enforcement of signals by aggregates of several or many identical optical particles. In particular, in the case of NPs, it can be very sensible to apply ~1 micron or larger assemblies instead of many single small particles in order to get high spatial contrast. Such a contrast enhancement by particle assembling is of high interest for image-based sensing. The sensing efficiency can suffer from a reduced access of analyte molecules to the sensing NPs in the case of dense assemblies. An incorporation of the assembling NPs into an analyte-permeable matrix could be an elegant solution for this problem. Thus, sensor NPs can be packed relatively dense into a swellable or nanoporous matrix, forming a submicron- or micrometer-size sensor particle incorporating hundreds, thousands, or more of the small sensing NPs. Particles with a size of ~1 micron and larger can easily be resolved by microscopic imaging and can give high contrast in the case of incorporating fluorescent components.

The preparation of metal/polymer micro- and nanoparticles for SERS sensing is a typical case for the desire for getting high sensor signals from a restricted small area. SNPs are best suited for this purpose. Despite silver, GNPs are used, too, in composite particles for SERS sensing.

A second motivation for assembling can be given by a combination of sensor signals. Several different types of NPs can be assembled in order to combine their signals locally. This can be of interest in the case of spatially resolved two- or multichannel measurements. The incorporation of particles contributing with an analyte-independent fluorescence intensity can be used for generation of local reference values for quantitative measurements.

The combination of oxygen and temperature detection is a typical example of a composite NP for dual sensing. This combined measurement succeeded by the application of a core/shell particle (Sung and Lo, 2012): The core material of CdSe-doped silica showed a temperature-dependent shift of 0.095 nm/K in the fluorescence maximum. The core contained a platinum complex with a characteristic phosphorescence that is quenched in dependence on the oxygen concentration.

Further, the embedding of groups of sensor NPs in larger sensor particles can be used for controlling the analyte access to the sensing NPs by the pore sizes and chemical properties of the embedding matrix. The solvent interaction with the matrix material, electrical charging by suitable functional groups and pH, lipophilicity, or hydrophilicity , as well as the molecular mobility of the matrix, can have a strong effect on the selective permeability of the matrix and can promote or suppress the transport of analyte molecules through the sensor matrix.

The combination of different NPs inside a microparticle can also be interesting for labeling purposes. In particular if several sensor particles are applied in parallel, it is very important to have a possibility to distinguish the different particles by optical labeling. Therefore, one type of NPs, whose optical properties are noninfluenced by the analyte, can be used for labeling and a second type for sensing. A typical construction could be a core/ shell microparticle incorporating labeling NPs or dyes inside a nonswellable and nonpermeable core and possessing a swellable and analyte-permeable shell in which the active sensor particles are incorporated. Advantageously, the permeable shell is not only accessible for the analyte but also has an extracting effect in order to enhance the analyte concentration inside the shell in comparison with the analyte liquid itself.

Chapter 4

Particle-Based Longer-Wavelength Electromagnetic, Magnetic, and Electrostatic Transduction

4.1 Infrared Readout

Infrared (IR) spectroscopy is estimated as to be a powerful tool for molecular analytics because of the high wavelength specificity that can be achieved. Many functional groups and substances can be identified by their specific fingerprint signals. This high specificity is due to the low bandwidth of electromagnetic resonances in the middle-infrared (MIR) region.

This high analytical power of IR analytics can be used for particle-based sensing if their IR spectra become changed by interaction with an analyte. This can be the case if new IR resonances are formed by a reaction of analyte components with the matrix of the particle. The effect can be particularly strong if an accumulation of these analyte components by the particle matrix takes place.

A nonreversible chemical reaction between particle matrix and analyte is not required in all cases for achieving an IR signal. Even fast equilibrium reactions can result in significant of IR responses from the particle. Such effects are observed, for example, in the case of protonation and deprotonation reactions, indicating changes in

Mobile Microspies: Particles for Sensing and Communication
Michael Köhler
Copyright © 2019 Pan Stanford Publishing Pte. Ltd.
ISBN 978-981-4800-14-3 (Hardcover), 978-0-429-44856-0 (eBook)
www.panstanford.com

the pH of the analyte medium. In addition, a shift in the bond strength due to surrounding effects could be detectable by IR measurements. Such effects are frequently found in solvent or ligand exchange reactions that can also be regarded as more or less reversible.

4.2 Particle-Supported Signal Transduction by Micro- and Radiowaves

In contrast to optical resonances in the ultraviolet-visible (UV-Vis) range, single-photon processes cannot be used for sensing in the microwave (MW) and in the radiowave (RW) range due to the low energies of the photons, which is much smaller than the thermal energies at room temperature:

$$h \times \nu_{MW/RW} << E_{th} = k_B \times T \qquad (4.1)$$

But the local absorption of MWs and RWs can result in a specific and significant local heating effect if the absorption of electromagnetic energy by the particle is higher than the absorption of the surrounding medium. This is the case if particles containing ions and/or dipole molecules are embedded in nonpolar solvents. A typical example would be the suspension of water-swellable polymer microparticles in a nonpolar organic liquid or the application of such particles at the interface between a nonpolar organic liquid and low-absorbing solids or gases.

In these cases, the presence of the microparticles causes a local change of the refractive index of the surrounding medium due to the local thermal expansion in cause of a MW-induced heating of the particle and the diffusive transfer of the evolved heat from the particle into its close neighborhood. Even small amounts of heat and small changes in the refractive index of the medium can be read out very sensitively be use of the *thermal lens effect*. Even a small gradient of the refractive index can be read out by detecting small deflections of a focused laser beam, for example.

The strength of the thermal lens effect is dependent on the intensity of incident MW or radio radiation and on the size of particle and the difference between the specific absorption by the particle and the medium. This last-mentioned influence can be used for getting quantitative data from the particle itself. At constant power of the incoming radiation and constant particle size, the optical lens effect can be taken as a measure of the particle composition.

4.3   Electrochemical Sensing and Electrical Signal Transfer

Besides optical properties, nanoparticles (NPs) are also suitable for applications with electrical or electrochemical signal transduction. A direct change in electrical charging, the modulation of voltages and electrical currents is particularly attractive for sensing because electrical signals can easily be converted by electrical transducers and taken up by electronic devices. The primary transduction by electrical sensing can be based on different effects:

- A change in the direct current (DC) conductivity of the medium
- A change in the electrical permittivity, respectively, in the alternating current (AC) conductivity
- A change in the mobility, resp. the electrophorectic sensitivity of charged particles
- A change in the zeta potential of particles by shifts in the particle netto charge
- Changes in electrochemical activities measured via an electrochemical mediator

In addition, sensing can be based on the optical readout of electrical effects, for example:

- Changes in spectral properties due to electrochemical reactions
- Coagulation of micro- or NPs due to reaction-induced charging or de-charging
- Electrically or electrochemically induced adsorption or desorption of particles at surfaces

The enhancement of electrical conductivity by incorporating electrically charged components is the simplest electrical effect that can be used. One possibility is the labeling with electrically charged metal or semiconductor particles. Another possibility is the selective incorporation or adsorption of cations or anions by the sensor particles. Whereas small ions have only low effects in many cases, very strong effects can be realized by the deposition of polyionic macromolecules at the surface of sensor particles (Fig. 4.1). These interactions can significantly alter the electrophoretic behavior of the particles.

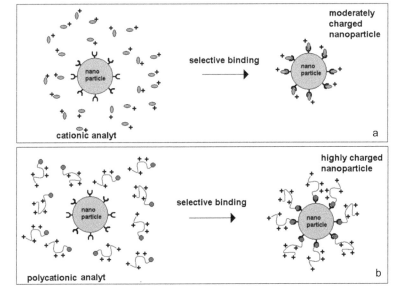

Figure 4.1 Generation of highly electrically charged nanoparticles by surface binding of polyionic macromolecules.

Particles can also be used for signal generation by impedance measurements. This type of measurement is based on a difference in the permittivity of particles and surrounding medium. Selective sensing occurs if functionalized particles would be trapped in a cavity, in a capillary slit, or at an interface. These binding events can then be read out by an AC because they are modifying the AC resistance due to a changed electrical capacitance. In particular in the case of an aqueous solution, the effects of impedance changing are large, for example, if polymer particles are used, because the permittivity of water is much higher than the permittivity of the polymer materials. Both single microparticles as well as groups of NPs can be applied for bead-based impedance sensing. The number and size of required particles depend on the geometry of applied measurement electrodes on the achieved permittivity differences of the materials and of the sensitivity of the electronic system, irrespective of the background noise. Impedance measurements are also applicable for the characterization of particles, cell content, or particle content inside microfluid segments (Fig. 4.2).

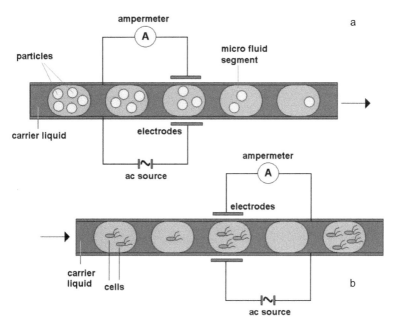

Figure 4.2 Detection of microparticles and cells inside microfluid segments by impedance measurements.

Electrically charged particles are accelerated in electrical fields as atoms or molecules are doing. In solution, an electrophoretic motion results, which depends on the applied voltage, on the size and charge of particles, and on the viscosity of the surrounding medium (Fig. 4.3).

Any change in the charge and size of particles results in a variation in the strength of the electrophoretic effect. Thus, binding or aggregation events or changes in electrical charge by ion adsorption or desorption by electrochemical reactions can be detected by changes in the particle mobility (Fig. 4.4). Vice versa, charged particles can also be used for detecting changes in the viscosity or porosity of the surrounding medium, which can be identified by changes in the response of the particles on an applied electrical field, too.

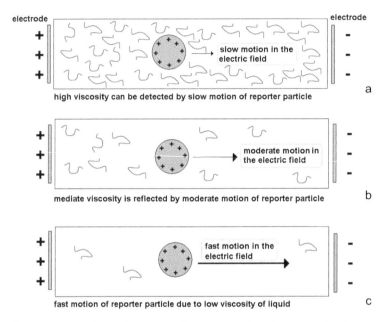

Figure 4.3 Readout of viscosity changes in a liquid medium by changing concentration of dissolved macromolecules using the effect of viscosity on electrophoretic actuation of sensor particles.

A change in the electrical charge of particles can directly be used for indication of a surface reaction with charged molecules. This effect is large if a lot of charged small molecules are binding or a higher-charged macromolecule—a polyionic macromolecule—is attached. The binding of DNA by hybridization is a typical example of such an effect.

The charging of particles can not only be read out by measuring the directed electrophoretic motion of particles in a constant electrical field but can be characterized by application of an alternating field. In this case, the oscillation amplitude reports about the charging of the particle. An optical readout of this motion by laser beams is applied, for example, in optical zeta potential measurements of ensembles of dispersed particles.

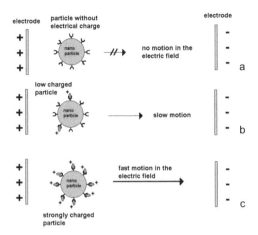

Figure 4.4 Nanoparticle-based measurement of charged analytes by modulation of electrophoretic forces due to selective binding of the charged molecules on the surface of functionalized nanoparticles.

In the case of metal NPs, the electrochemical potential is mostly dominating the particle charge in colloidal solutions. Zeta potential measurements of these particles allow to detecting changes in the electrochemical potential of the particles. Metal NPs, which are dispersed in an electrolyte, can be always regarded as small open-circuit electrodes. Adsorption or release of ions and ligands causes the electrical charging of the particles (Fig. 4.5). Besides the interaction of the particle with metal ions of the particle material itself, other electrochemical processes can contribute to the total electrochemical potential, too. In these cases, the NP has to be regarded as a mixed electrode and each change in the concentration of an electrochemical active species would cause a shift in the readable particle potential.

In some cases ongoing chemical reactions are not directly interfering with dispersed NPs. But these chemical processes can be connected with electrochemical processes on a particle surface by a species that can be reduced and re-oxidized by reversible processes. Such substances can act as an electrochemical mediator between the redox chemical state of an analyte in the medium and the charging of metal NPs (Fig. 4.6).

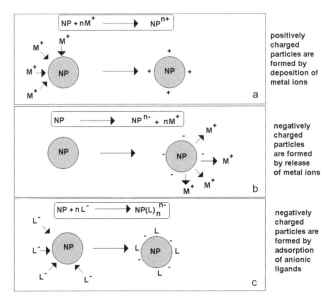

Figure 4.5 Electrical charging of metal nanoparticles in colloidal solution by adsorption or release of metal ions and interaction with electrically charged ligands.

Besides the electrical readout of electrical or electrochemical processes on a sensor particle, optical changes can be induced by electrochemical reactions, too. Dispersed metal NPs (so-called plasmonic particles), which are marked by a specific electromagnetic resonance, react with a shift in the absorption peak and oscillator strength if their size, shape, or surface charge is changed. This can be detected by measuring the intensity of scattered or transmitted light or by measuring the shift of the so-called plasmon resonance peak.

The spectral properties of metal NPs are mostly drastically changed if particles are not only modified gradually by charging, decharging, or chemical surface reactions but also aggregated or coagulated. The coupling of the electronic systems of particles, which are coming in contact, leads to a strong shift in the color of their colloidal solutions. Typically, a bathochromic shift and an enlargement of the absorption bandwidth take place. In the case of the formation of larger aggregates, the appearance of broad absorbance is observed that can be easily recognized by the substitution of red, green, and blue colors of the colloidal liquid by gray or black scattered light. Such aggregation or coagulation is frequently caused by a de-charging of electrostatically stabilized metal colloids due to chemical surface reactions or a charge compensating adsorption of ions. Well-detectable coagulation events can also be initiated by the addition of antagonistically charged particles to a charge-stabilized colloidal system.

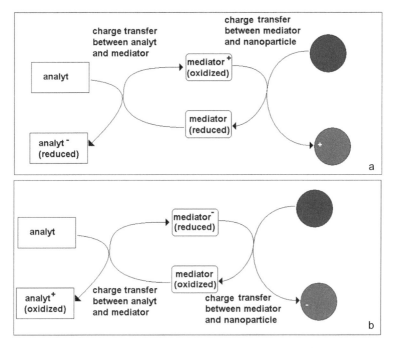

Figure 4.6 Detection of redox-active analytes by charge transfer mechanisms to colloidal nanoparticles using a molecular redox mediator.

Since the adsorption or desorption of particles from a surface can be recognized by optical detection, electrically or electrochemically induced surface interactions can also be applied for particle-based sensing. It allows the detection of chemical processes if they are modifying their electrical state and the electrically controlled interaction of particles with surfaces. This strategy is suitable in connection with transparent electrodes, for example, that allow to modulating the particle–surface interaction by shifting the electrode potential.

4.4 Signal Transduction by Magnetic Particles

Magnetic microparticles are well suited for application in transduction chains for contact-less measuring of small mechanical forces locally. The particle transfers a local direct mechanical interaction to the noncontact force sensing of an outside electromagnetic or magnetomechanical system. This strategy was applied, for example, in the characterization of contracting bead-coupled single cardiac myocytes by an external magnetic field. Contraction forces in the order of magnitude of 5 µN could be measured by this method (Yin et al., 2005).

Magnetic microparticles are well suited for all kinds of force-compensation-based measurements. Any force on such a particle inside a liquid or gaseous system can be compensated by an outside magnetic field. This concerns earth gravitation, streaming of liquid, or even local electrostatic fields accelerating the microparticle in the case of an electrical charge. An important advantage of magnetic fields is the fact that magnetic fields are less shielded by conductive media as electrolytes, semiconductors, or metals, on the one hand. On the other hand, the motion of magnetic microparticles can be traced by optical methods, for example, in microfluidic systems, if the walls are transparent. Particle imaging allows also the simultaneous tracing of ensembles of several or many magnetic particles. Thus velocity distributions or even force distributions under complex streaming conditions can be characterized.

The catching of substances by magnetic particles and the subsequent separation or particles from a liquid medium are the most important applications of magnetic particles for information transfer from a liquid system. In this case, the particles are not working as a transducer directly but are used as a tool for recognizing and extraction of substances or even larger objects like viruses or cells from the liquid media. The magnetic properties of the particles are only required for a convenient separation procedure and convenient rinsing or otherwise processing of the trapped substances. The specificity is achieved by specific recognition and binding elements that are immobilized at the particle surface. There, the whole spectrum of specific chemical binding can be applied, for example, the binding of metal ions and small cationic molecules by immobilized crown ethers or cryptands or the biomolecular binding by DNA–DNA hybridization or the application of immobilized antibodies for specific fishing for proteins, viruses, and whole cells.

Iron oxide NPs with superparamagnetic optical properties (also called SPIONs) became particularly important. These particles are of particular interest for combination of sensing, monitoring, and controlled interactions and are used, in particular, for operations combining biomolecular recognition, tissue characterization, and drug release in medical applications (see Section 6.3).

4.5   Ultrasound-Supported Sensing by Particles

The affinity of molecular-functionalized particles can used for the enforcement of optical effects of the reaction of analyte molecules with immobilized reagent molecules. Particles can supply a significant contribution to optical properties of interfaces as reflectivity, light scattering, and surface plasmon resonances, as well as electrical conductivity and impedance. The selectivity of particle binding can be improved considerably by the application of ultrasound (Glynne-Jones et al., 2010). Probably, the effect of ultrasound is comparable to moderate thermal activation that supports the differentiation between slightly stronger and slightly weaker bonds.

Chapter 5

Construction Types and Preparation of Sensor Particles

5.1 One-Component Responsive Nanoparticles

Nanoparticles (NPs) of a single material which can respond by changes in physical properties on environmental effects represent the simplest variant of sensor NPs. Among different materials, gold nanoparticles (GNPs) are of big interest because of their high chemical stability, on the one hand, and their responsivity in the optical and electronic properties on changes in the surrounding medium and at its surface, on the other hand. GNPs are in use for different transduction principles using changes in refractive index, in electrical conductivity, by electrochemical activity, by catalysis, and by mass changes in connection with piezoelectric oscillators (Y. Li et al., 2010).

Particle "spies" consisting of one single component only have no special construction. Their ability for readout and transmission of information is due to a specific material property. What makes them act as a transducer is the response of the material on a change of the surrounding medium in a way that can be detected from outside, preferably by a noncontact readout method. Some examples for measuring temperature and mechanical effects by use of optically active inorganic sensor NPs are given in Table 5.1.

Mobile Microspies: Particles for Sensing and Communication
Michael Köhler
Copyright © 2019 Pan Stanford Publishing Pte. Ltd.
ISBN 978-981-4800-14-3 (Hardcover), 978-0-429-44856-0 (eBook)
www.panstanford.com

Table 5.1 Examples of in situ transduction of physical signals by special inorganic sensor particles

Measured parameter	Resolution	Particle type	Measurement object	Ref.
Temperature	0.2 K	Quantum dot/ quantum rod nanothermometers	HeLa cells, NIH 3T3 cells	Albers et al., 2012
Temperature	–	Rare-earth up-converting NPs	HeLa cells	B. Chen et al., 2013
Temperature	0.7 K	Nd^{3+}-doped LaF_3 NPs	Tissue	Rocha et al., 2016
Mechanical forces	$(\mu N - nN)$	Rare-earth up-converting NPs	Material stress, tissue	Lay et al., 2017

A very simple case of simple single-material particle sensor type is a swellable polymer particle. The swelling can be recognized by optical imaging if a significant volume increase takes place (Fig. 5.1). Alternatively, in a flow system, the increase of particle diameter can also be recognized by the residence time in a light beam, for example, by a microphotometric arrangement. Such a sensor can be applied for distinguishing different solvents or for the determination of the quantitative composition of a binary solvent mixture. The swellability in water is strongly affected by the ability to form hydrogen bonds and by the presence of ions that make a strong ion–dipole interaction with water molecules. This way, the swellability of a particle can also be applied for the detection of ions. The compensation of electrical charges in the matrix by ions from the environment can lower the matrix hydrophilicity and, therefore, its swellability. An increase of electrical charge density by a reaction of the matrix material with ionic partners causes an increase in swellability by water.

The introduction of chemical recognition functions can make one-component particles more intelligent. A permeable polymer or gel particle that changes its optical properties by the interaction with certain analytes can represent a very useful particle sensor. A change in absorbance or fluorescence is the easiest way for a contact-free readout of a reaction of the particle.

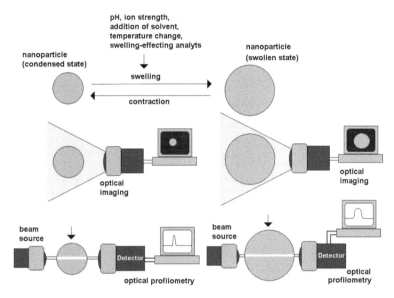

Figure 5.1 Optical readout of swellability-based sensing with polymer particles.

Specific sensitivities can be achieved by molecular imprinting. Imprinted materials are generated by formation of a comparatively stiff but swellable matrix-embedding analyte molecules. After selective dissolution of the analyte molecules, nanoholes are obtained that match the shape and size of analyte molecules and act as highly specific recognition sites. Molecular-imprinted polymer (MIP) materials are formed by polymerization and cross-linking in the presence of the analyte. This technique can also be transferred to the generation of MIP NPs (Wackerlig and Lieberzeit, 2015; Weber et al., 2018). The principle of formation of molecular-imprinted sensor particles for detection of special analytes is shown in Fig. 5.2. The analyte is used for the generation of characteristic holes inside the polymer matrix of the particle. The particle and the shape of the holes are stabilized by chemical cross-linking. The binding of analytes can be detected either by selectively proving the accumulation of analytes inside the particles or by the change of physical properties of particles by the loading with analytes, in general.

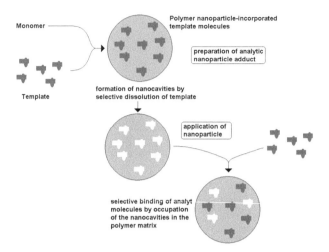

Figure 5.2 Principle of generation of molecular-imprinted sensor particles.

Supermolecular host structures have analog functions as MIP recognition sites. They are also suitable for a more or less specific binding of analyte molecules and can be combined with the polymer matrix of a sensor particle (Gontero et al., 2017). Besides crown ethers and cryptandes, calixarenes are frequently used as host structures for analyte recognition. Calixaren-functionalized gold and silver NPs have been developed, for example, for the detection of different amino acids like arginine, aspartic acid, glutamic acid, glycine, leucine, methionine, histidine, tryptophane, and histidine (Makwana et al., 2017).

A particle matrix with an immobilized pH-dependent fluorescence-changing dye can act as an optical pH sensor. Fluorescence transduction is a suitable principle for larger and smaller particles because of its high sensitivity. Alternatively, an immobilized indicator dye, which changes its color in dependence on pH, can be used if the dye concentration can be high enough and the particle is not too small. Lambert–Beers law explains that for photometric detection with an absorbance of about 0.1 the particle should have a diameter of at least 5 μm, if the molar extinction coefficient is about 20,000 L/(mol.cm) and a chromophore concentration of 10 mM can be achieved.

Fluorescent pH sensor particles have been synthesized by the group of Nagl and Wolfbeis (2007). These particles have typical diameters in the middle to lower micrometer range. They are well applicable in microfluidic systems. They are particularly suitable for measuring pH changes in aqueous solutions due to the metabolic activities of microorganisms or other cells because they are marked by a strong pH dependence of fluorescence quantum yield in the physiologically particular relevant pH range of 5 and 9. Thus, these particles can be calibrated for a rather accurate quantitative determination of pH in microliter or nanoliter droplets, which means in a very small volume. In a similar way, oxygen sensors can be realized as one-component sensor particles when a triplet dye is added. Particular sensitive oxygen sensors have been prepared by phosphorescence dyes incorporated in the oxygen-permeable polymer matrix (see Section 3.2.2).

Besides polymer particles, assembled particles of smaller organic molecules can be used as sensors. Thus, the fluorescence of tryptophane-phenylalanine dipeptide NPs is dependent on the rigidity of the matrix and the environment. These particles can be functionalized, for example, by aptamers and used for selective detection of different biomolecules by a shift either in the fluorescence wavelength or in the fluorescence quantum yield (Z. Fan et al., 2016).

A special case of fluorescent polymer nanoparticles (FPNs) is represented by micellar assemblies. Micelles are spontaneously formed if the critical concentration of an amphiphilic substance—the critical micelle concentration (CMC)—in a solution is exceeded. In contrast to NPs formed by polymerization, the micelle formation is a reversible process, and micelles are dissolved spontaneously if the solution concentration of the micelle-forming amphiphilic substances is lowered. Responsive micelles that might be usable for local sensing could be chemically sensitive by significant shifts in the CMC. Otherwise, micelles with very low CMCs are suitable for labeling supporting microscopic imaging (Li et al., 2015).

5.2 Surface-Functionalized Micro- and Nanosensor Particles

5.2.1 Surface Functionalization for Molecular Recognition Sites

Molecular recognition elements and molecular functions for signal transduction can also be placed on the surface of particles. The immobilization strategies for the functional elements are analogous to the established strategies in biosensors. The only difference is found in the fact that not a selected part of a fixed sensor surface but the surface of a particle is functionalized.

The surface functionalization can be performed in suspensions of micro- and nanoparticles as well as in colloidal solutions of NPs. In the last mentioned case, one has to take care that the colloidal state is not destabilized by the surface-modifying reactions. The danger of a coagulation of particles is particularly high if the surface reactions are lowering the electrical charging of particles by de-charging of the functional groups.

Examples of useful techniques for functionalization of micro- and nanoparticles for sensing are given here:

- Unspecific adsorption
- Electrostatic binding
- DNA hybridization
- Antibody coupling
- Coordinative coupling
- Covalent linking
- Surface polymerization

The unspecific hydrophobic bonding can already be caused by the weakest possible and general available van der Waals forces, which are a universal attractive interaction between all atoms and molecules. The strength of van der Waals interaction increases with the number of surface atoms of a molecule and is therefore determined by its size. Small molecules that are adsorbed by van der Waals forces only can quickly escape from a surface. In contrast, macromolecules as synthetic or biomacromolecules can be adsorbed and immobilized at a surface by their relatively strong van der Waals forces. In aqueous environment, polar interactions such as dipole–ion forces, dipole–dipole forces, and hydrogen bridges mostly dominate the molecular interaction. The van der Waals forces are important for interaction of molecules without protic groups and without dipole character. In particular in aqueous environment, such molecules show a preferential binding to other nonpolar molecules or particles by the unspecific attractive forces between their surfaces. These attractive forces have the character of a hydrophobic or lipophilic interaction. They can be applied for assembling of nonpolar molecules in a polar medium, for particle formation by molecular aggregation, and also for their immobilization on nonpolar surface areas of nano- or microparticles.

Immobilization of recognition or transducer molecules by electrostatic forces demands for permanent antagonistic charges or by dipole–dipole interaction. In general, dipole–dipole interactions are significant weaker than the direct electrostatic attractions between positive and negative electrical charges. Thus, ions can be much easily and strongly immobilized by countercharges on particles than dipole molecules by surface dipoles. Dipole–dipole interactions might become sufficiently stable for immobilization if several dipole groups are involved in a collaborative interaction. These interactions have the character of polyvalent bonds and play an important role in the adsorption and immobilization of polar macromolecules. In an aqueous environment, all polar interacting molecules are always in competition with water, which is a strongly solvating polar interaction partner, and always bond in the coordination sphere of each polar molecule and ion. Therefore entropic effects are very important for polar binding in aqueous solutions. Consequently, molecules with several ionic or functional groups have much stronger binding to charged particle surfaces than small ionic or polar molecules. A very high binding efficiency is observed in the electrostatic binding of polyionic macromolecules, which is used in the preparation of polyelectrolyte multilayer (PEM) films and the immobilization of sensor molecules in such films.

Polyvalent bonding is also important for the immobilization by hydrogen bridges. Since a single hydrogen bridge is too weak for sufficient strong binding, groups of hydrogen bonds are well suited for the immobilization of molecules and even larger recognition molecules on the surface of sensor particles. The cooperative and highly specific bonding of single-stranded DNA by hybridization with complementary oligonucleotide sequences is largely used for molecular coupling and immobilization as well as for recognition and sensing itself.

Besides DNA, proteins can also be used for tight attachment of molecules on the particle surface. In particular, the binding of antigen molecules by their specifically binding antibodies is a usual immobilization strategy. The antigen–antibody coupling can be performed, for example, by surface-attached streptavidin that is selectively recognizing and strongly bonding the small molecules biotin.

Besides biomolecular strategies, even inorganic interactions are usable for immobilization procedures on particle surfaces, among them being coordinative coupling (formation of complex compounds). In this method, the formation of coordinative bonds between at least two ligands and a metal ion is used. A frequently used system is the complex formation of a nickel complex by oligohistidines that form strong bindings with nickel ions (*nickel his-tag*). The presence of suitable metal ions is mandatory for this type of coupling because the metal ions are required as central units in the coupling complex compounds, always. For the coupling, chelate ligands forming particularly stable complexes as the nickel-hist-tag have to be applied in order to get a stable attachment. Many coordinative bonds are comparatively weak and have to be regarded as reversible interactions.

Molecular functionalization can be used for biomolecular recognition as well as for measuring inorganic components. There exists a large spectrum of possibilities for chemical or biomolecular functionalization of particle surfaces (Fig. 5.3). Some examples of using polymer particles with molecular bonding structures for recognition of metal ions are given in Table 5.2.

An alternative is the application of covalent linkers. Covalent bonds are much more stable than coordinative bonds, in general, and have to be regarded as irreversibly formed, in most cases. In many cases, the recognition molecules can directly immobilized by covalent coupling on the surface of sensor particles by suitable chemical functions. In other cases, a special linker molecule is used for connecting the molecule with the particle surface. Sometimes, in addition, a spacer unit is applied to attach the recognition molecule in a certain distance to the surface or to ensure a sufficient high local molecular mobility. There are plenty of possibilities for covalent coupling in principle. But some of them are particularly convenient and frequently applied. Among them the coupling by forming silane or siloxane groups (*silan coupling*) and the azomethin coupling by reaction between an aldehyde and an amino group count as the most favored.

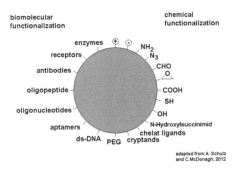

Figure 5.3 Variants of chemical and biomolecular functionalization of micro- and nanoparticles for sensing applications (Schulz and McDonagh, 2012).

A special variant of covalent coupling is given by the in situ immobilization in a copolymerization process. Different recognition molecules can be included in surface polymerization and form finally a surface film consisting of a 3D molecular network. This immobilization can be made particularly reliable and efficient if the recognition molecules for immobilization are equipped with a special polymerizable coupling function, for example, an acrylate group. In some cases the immobilization of molecules on particle surfaces succeeds also without direct covalent bonding of the molecules, but by catching the molecules due to their unspecific surface interactions during the surface polymerization. As a result, the recognition molecules are trapped in the polymer matrix. This trapping can be very efficient if the recognition molecules are comparatively large and the 3D bond network, which had been formed in the polymerization process, is dense. Otherwise, there exists the risk of successive loss of recognition molecules by dissolving from the polymer matrix.

Table 5.2 Examples of molecular-functionalized polymer microsensor particles

Matrix material	Size	Function	Transduction	Preparation	Ref.
PVC	10 μm	Na sensing	Fluorescence	Precipitation in droplets	Tsagkatakis et al., 2001
MMA-DMA-Copol.	10 μm	Cation sensing	By chromo-ionophore	Polymerization in droplets	Peper et al., 2003
PVC (+ plasticizer)	~20 μm	Potassium sensing	By chromo-ionophore	Emulsification and curing	Ye et al., 2007
Polystyrene	~2 μm	Cu^{2+} sensing	Spiropyran interaction	Surface functionalization by spiropyranes	Scarmagnani et al., 2008

5.2.2   Fluorescent Silica Particles for Chemical and Biomolecular Sensing

The fluorescent silica NPs represent a special type of fluorescence probe. They have a large spectrum of possible applications and can be generated by a comparatively convenient general approach (Schulz and McDonagh, 2012; Korzeniowska et al., 2013). The development of functionalized fluorescent silica NPs had established a promising new platform for a large spectrum of sensing applications. It is usable for staining in microscopy in general as well as for special applications for characterization of different cells, tissues, and whole small animals (K. Wang et al., 2013). This particle type represents a nice example of the creation of a universal concept and basic tools by a spatial separation of functions: The fluorescence dye–loaded silica matrix represents the universal part. It is responsible for the optical readout and for the spatial resolution of measurements. The functionalization of the particle surface represents the flexible part. It is responsible for the specificity of interaction, for the efficient recognition of target molecules, and for the generation of an analyte-dependent signal.

Typically, silica NPs are formed by the hydrolysis of a silica precursor like tetraethylorthosilicate (TEOS) and the condensation of the intermediately formed silica acid (Bagwe et al., 2004; Van Blaaderen and Vrij, 1992). Advantageously, this condensation reaction proceeds together with linker molecules like 3-amino-propyltriethoxysilane (APTES) in order to introduce binding groups for fluorescence dyes. Functionalized dyes can then be covalently coupled by reaction with amino groups, for example, by N-hydroxysuccinimide (NHS)-ester or isothiocyanate functions. In result, silica NPs with immobilized fluorescence dyes are obtained.

These particles have to be functionalized on their surface in order to achieve a specific binding affinity for the target analyte. The amino groups of the introduced linker in the silica matrix are used for this coupling, too. The amino function can be converted to other binding functions by reaction with a secondary linker, for example, for the introduction of thiol, carboxyl, aldehyde, or epoxy function on the particle surface. These functions allow then the immobilization of highly specific recognition molecules for chemical or biomolecular sensing as single-stranded DNA, aptamers, oligopeptides, proteins, antibodies, cryptands, or other complex compounds for interactions with smaller chemical objects. A simple mechanism for a fluorescence switch-on mechanism using the host–guest principle is shown in Fig. 5.4. In this case, the incorporation of analyte molecules into matching molecular macrocycles or cryptands, which are immobilized on fluorescent NPs, leads to a reduction of electronic quenching of the particles.

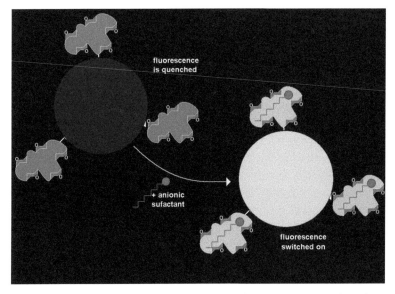

Figure 5.4 Example of molecular sensing by fluorescence quenching host compounds immobilized on the surface of fluorescent nanoparticles; the fluorescence is recovered by reaction of the guest molecules (analyte) with the surface-attached host molecules.

The coupling of the recognition molecules with fluorescent silica NPs instead of a direct chemical coupling with molecular fluorophores offers two important advantages: First, the fluorescence dyes are, in general, more stable in the silica environment than in solution. Second, the incorporation of many dye molecules in an NP leads to a strongly improved contrast of fluorescence images that is very important for measuring local analyte concentrations and for the detecting concentration gradients.

The different sensing functions are realized by the specific coupled recognition molecules: Ruthenium complexes (Xu et al.,

2001) and platinum complexes (Koo et al., 2004) had been introduced for oxygen sensing and pH-sensitive dyes for pH measurements (Burns et al., 2006). Despite oxygen and pH, metal ions are also of interest for physiological characterization of tissues and cells and can be detected by silica NPs if corresponding metal-sensitive indicator dyes are present on the particle surface or in the pores of a mesoporous silica particle matrix. For example, zinc ions (Sarkar et al., 2009; He et al., 2010) as well as copper ions (Seo et al., 2010) had been measured in HeLa cells by the application of such silica NPs. The characterization of cancer cells with silica NPs succeeded by immobilization of specific antibodies (Ow et al., 2005; Smith et al., 2007; Peng et al., 2007; Cheng et al., 2010).

5.2.3 Fluorescent Carbon Nanoparticles

In the past, a large spectrum of different carbon NPs has been found, which can be produced from a lot of different precursors, whereby very different preparation methods are applied. Fluorescent carbon nanoparticles (FCNs) can be formed by different degradation processes, including electrochemical conversion (H. Li et al., 2010) or laser ablation of graphite (Suda et al., 2002) and pyrolysis of organic compounds (Fortunato et al., 2010; Yang et al., 2011), and polymers (Liu et al., 2009; Jiang et al., 2012). The typical sizes are in the lower nanometer range (up to about 10 nm).

Fluorescence is dependent on the preparation conditions and the content of heteroatoms. Thiamine (vitamin B_1) can act as an interesting precursor for FCN preparation. Carbon particles with high fluorescence can be prepared by thermal degradation of this precursor in the presence of sodium triphosphate (Bhunia et al., 2014). The obtained particles are stable and well suited for the staining of HeLa cells. The reported results promise that FCNs might also become interesting for mobile sensor NPs after functionalization or by co-immobilization with recognition molecules on larger NPs.

5.2.4 Up-Converting Nanoparticles for Biolabeling and Sensing

Up-converting NPs or up-converting nanophosphors (UCNPs) are NPs that are able to absorb photons of lower energy and emit photons of higher energy. Typically two or several photons of far-red or near-infrared (NIR) radiation are absorbed and converted into short-wavelength photons of visible light or near-ultraviolet (UV) photons. High efficiency of such conversion processes was found in the case of some rare-earth materials, for example, on the basis of yttrium and erbium fluorides (Wang and Liu, 2008; F. Wang et al., 2010).

Such particles show temperature-dependent luminescence intensity. Therefore, they had been applied for nanoscale temperature sensors and for combinations of cell staining and localized temperature readout (Chen et al., 2013).

The fluorescence of doped $NaYF_4$ NPs is also affected by mechanical forces. Different direction and strength of changes in luminescence intensity in dependence on pressure have been observed for different crystal structures. The pressure-dependent luminescence intensity can be used for a local and spatial-resolved characterization of pressure and tensions inside microscopic objects as materials, cells, and tissues (Lay et al., 2017).

5.3 Swellable and Gel-Like Molecular-Doped Polymer Micro- and Nanoparticles

Gel-like particles can either be formed by the swelling of a preformed particle with a partially cross-linked matrix or by cross-linking polymerization of dissolved monomers under in situ forming of particles in the gel-like state. Sensor functions as molecular recognition and molecular transduction can be implemented afterward if the matrix is permeable enough for diffusion of the reaction components or by simultaneous incorporation during the polymerization process itself.

A cross-linked polyacrylamide matrix is a typical case for the microfluidics preparation of gel microparticles. Typically, an aqueous solution of acrylamide, low addition of bisacrylamide for cross-linking, and a photoinitiator form the reaction mixture (Serra and Chang, 2008). The solid content of the finally obtained gel particles depends on the concentration of the monomer mixture inside the reaction solution. The solid content can be enhanced by drying the particles after polymerization is completed. Despite the fact that different reactants are mixed for generation of these particles, they are homogeneous and are regarded, therefore, as monocomponent particles.

Ion-sensitive sensor particles have been obtained from emulsions by polymerization of droplets of monomer mixtures containing chromo-ionophoric substances. Such particles are applicable for sensing of protons or potassium ions and can also be applied for the measurement of some anions by proton–anion co-extraction (Peper et al., 2003). Such particles have to be read out by the concentration-dependent change of their absorption. The sensor particles are also called microsphere-based ion-selective optrodes (Ye at al., 2007).

5.4 Two-Phase Composite Particles

The construction of two-phase sensor particles can be motivated either by an improvement of handling or by an enhancement of sensitivity or specificity of sensing. In both cases, it is of interest to use smaller particles as the sensing elements, whereby the signal transduction profits from the high specific surface and the short transport paths in the small particles.

It is crucial that the analyte molecules have access to the small sensor particles. Thus, they have either to be placed at the surface of the larger particles or to be incorporated inside a matrix that is permeable for the analyte. Nanoporous or gel-like materials are suitable for the realization of high permeability in the NP-embedding matrix. The chemical properties of this matrix can also be adapted for enhancement of sensing specificity by promoting the transport of some components of a mixed analyte and to suppress the transport of other components. A matrix-dependent solubility can also be used for enhancing of sensitivity and selectivity of composite sensor particles with embedded NPs.

Referencing is a crucial problem for getting absolute values from measurement and to guarantee comparability of different measurement procedures. Composed sensor particles offer the possibility of particle internal referencing.

A core/shell silica NP was developed for measuring hypochlorite by a reference method (Zhang et al., 2015). Therefore, two different phosphorescent iridium complexes providing different colors of fluorescence light have been incorporated in the core phase and the shell of the silica NP. Whereas the fluorescence intensity of the complex in the shell is dependent on the hypochloride concentration, the fluorescence intensity of the complex in the core is independent on hypochlorite and can be used for reference.

Besides pure polymer particles with imprinted cavities, composite particles offer additional advantages for molecular sensing. A promising concept is the incorporation of transducing NPs into an imprinted polymer matrix of a microparticle. This construction connects a high density of binding sites with short distances between analyte molecules and nanotransducers inside the matrix. Incorporated noble metal NPs can be used by their plasmonic properties as well as for detection by surface-enhanced Raman spectroscopy (SERS), for example.

An interesting strategy for the realization of an imprinted matrix doped with plasmonic particles was demonstrated by Riskin et al. (2010). They incorporated gold nanoparticles (GNPs) by electropolymerization and cross-linking of thiol-substituted bis-anilines in the presence of the target analyte (templates). After dissolution of the template molecules, molecular wholes ("footprints") with the size and shape of the target molecules were obtained in the polymer matrix close to the co-embedded transducing metal NPs. The authors were able to generate sensor materials for stereo- and chiroselective recognition of small biomolecules as amino acids, for example.

5.5 Multicomponent and Hierarchically Constructed Sensor Particles

5.5.1 Bi- and Multicomponent Construction Strategies

The sensor performance of composed micro- and nanoparticles can further be improved by special compositions and constructions using more than two levels of spatial organization. There is a large variety of strategies for construction of complex composed micro- and nanoparticles. From the point of view of spatial organization they can be ordered by different types. Some examples are given here:

- Binary particles
  - Janus particles
  - Core/shell particles
  - Head/tail architectures
- Multicomponent particles
  - Globular aggregates
  - Chain-like aggregates
  - Branched aggregates
  - Dendritic aggregates
  - Smaller particles forming a shell around a larger particle
  - Smaller particles incorporated in the matrix of a larger particle
  - Smaller particles incorporated in the shell of a larger core/shell particle
  - Smaller particles incorporated in the core of a larger core/shell particle

Some of the aggregate structures can be formed spontaneously in the case of suitable chemical conditions, in situ during the formation of particles or after mixing or preformed particles, or by changing the solvation conditions and the conditions for stabilization of colloidal particle solutions. Well-defined nonglobular aggregates demand for special binding sites or for a special particle synthesis strategy. Microfluidic strategies are well suited for realizing Janus-type, core/shell or core/double-shell microparticles, necklaces, and other shapes (Serra and Chang, 2008; Serra et al., 2013a) and for controlling particle assembling for generation of functional microparticles (Bouquey et al., 2008).

Up to now, several types of bi- and multicomponent micro- and nanoparticles have been reported. Binary particles represent the simplest cases of composed particles. Globular aggregates are the simplest assemblies. In addition, there is a large spectrum of composed particles on the basis of different construction concepts. Some of them are shown in Fig. 5.5.

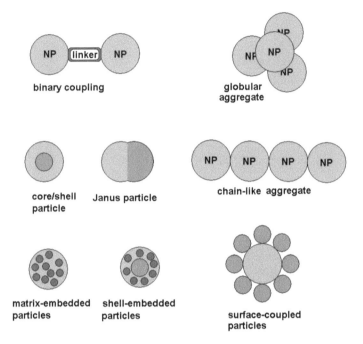

Figure 5.5 Examples of construction types of multicomponent sensor particles.

5.5.2 Nanoparticle-Doped Core/Shell Microparticles

An interesting strategy is to immobilize sensitive NPs inside the matrix of a shell. Thus, the access of analytes to the sensing particles is marked by reduced variation of the length of diffusion paths, and response times of the sensor particle are reduced. Such core/shell particles can be regarded in analogy to the deposition of an NP-containing sensing layer on a solid substrate.

5.5.3 Functionalized Microcapsules for Sensing

The construction of microcapsules uses the fundamental principle of a living cell. A self-forming shell of molecules separates an internal medium from the environment. There are several strategies for realizing such capsules, which have been developed for sensing as well as for other chemical applications. Vesicular structures are near to the construction of natural cells (Lensen et al., 2008). In these types of capsules amphiphilic molecules form closed molecular double layers separating the hydrophilic environment from an also hydrophilic internal medium. These cell-like capsules are named polymerosoms if amphiphilic macromolecules are applied for their formation. Amphiphilic block copolymers with a hydrophilic fraction between 25% and 40% are advantageously used in the polymerosoms. At higher hydrophilic contents, rod-like molecular aggregates of micelles are observed instead of capsules (Discher and Ahmed, 2006). Mostly, two-block copolymers containing polystyrene (PS) or polybutadiene for the hydrophobic part and poly(acrylic acid) or polyethyleneoxide for the hydrophobic part are applied. In alternative, amphiphilic blocks can also be constructed by polypeptides consisting of a block of amino acids with hydrophilic side groups and a second block of amino acids with hydrophobic side groups. The synthesis of hybrid macromolecules composed by peptide blocks and synthetic polymer blocks leads to so-called *peptosomes* (Kukula et al., 2002).

Oil-in-water microcapsules can easily be achieved by a phase separation strategy (Lensen et al., 2008). Therefore, an oil-in-water emulsion is generated in which the oil phase contains a dissolved polymer and a solvent mixture composed by a volatile solvent and a nonvolatile nonsolvent for the polymer. The capsules are formed by precipitation of the polymer near the water–oil interface due to the decreasing concentration of the polymer solvent during its evaporation. A similar method uses such an emulsion for the formation of a polymeric microcapsule by a combination of polymerization and precipitation at the interface. The typical approach is the application of a bi- or trifunctionalized reaction partner in the oil phase and a complementary bi- or trifunctionalized reactant in the water phase. The polymer material is then formed by direct interaction of both reactants and remains in the interface between both the liquid phases. Typically, reaction partners of high reactivity and affinity are applied. A convenient way is, for example, the application of a bifunctionalized carboxylic chloride in the organic phase and di- or triamine in the aqueous phase. The density of cross-linking of the matrix can be controlled by the ratio of bi- and trifunctionalized components in the monomer mixtures. The formation of capsules by aggregation or sintering of colloidal NPs instead of monomers results in so-called *colloidosomes* or *pickering emulsions* (Yow and Routh, 2006).

The formation of microcapsules can be achieved by use of the layer-by-layer (LBL) technology, too. This method was first developed for the formation of molecular films on larger solid surfaces by alternating deposition of polyanionic and polycationic macromolecules (Decher et al., 1992; Decher, 1997). The trick of this technology lies in the fact that polyionic macromolecules of one large excess charge are electrostatically bonded to an antagonistically charged solid surface, but they not only compensate the surface charge, but also overcompensate it, thus generating an opposite excess charge at the surface, which can now bond macromolecules of surface-antagonistic charging. Many alternating films of polycationic and polyanionic macromolecules can be deposited successively for forming a thicker molecular stack (LBL multilayer stack) by this way.

5.5.3.1 Formation principle of LBL-coated particles and LBL capsules

By modification of this method, the extended plane solid surface can be substituted by the surface of microparticles in order to equip them by an LBL multilayer stack. Such particles can be converted to a molecular capsule if their solid core is selectively dissolved (Fig. 5.6). During the deposition of the single layers of the macromolecular stack, it is possible to co-immobilize additive objects like sensor molecules of sensor NPs. Thus, microsensor capsules with shell-incorporated sensor NPs are formed (DelMercato et al., 2014). Multifunctional polymer microcapsules have been developed for different biotechnical and bioanalytical applications during the past years. Therefore, the integration of different transducing elements in the polymer shell was established (Sukhorukov et al., 2007), among them the incorporation of quantum dots, superparamagnetic NPs, and metallic NPs and the incorporation of enzymes and antibodies. The polyelectrolyte microcapsules (Zan et al., 2015) have a large spectrum of promising applications for drug release, for nanoreactors, and for sensing. They offer a lot of possibilities of constructing complex nano- and microparticle architectures (Ariga et al., 2012) and could become a basis for multisensing approaches.

Such capsule-type sensor particles can be used for highly specific catching of nucleic acids by hybridization with capsule-integrated DNA or of proteins by immobilized antibodies (Verma et al., 2016). Microcapsules made by polymerization of polyallylamine under embedding of silica NPs had been labeled by fluorescein isothiocyanate (FITC), resulting in pH-sensitive capsules that can be read out by fluorescence. These particles can be introduced into cells and used for pH sensing in the interior of HepG2 and endothelian cells (Zhang et al., 2014).

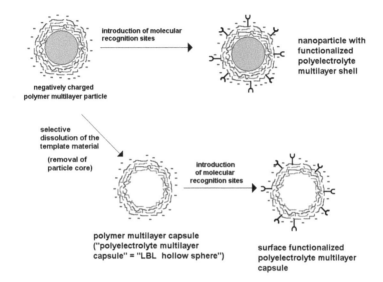

Figure 5.6 Formation principle of LBL core/shell particles and LBL microcapsules.

5.6 Special Microfluidic Techniques for Preparation of Sensor Particles

5.6.1 Particle Formation by Thermal Polymerization of Monomer Droplets

Microfluidics offers convenient methods for generating droplets of very regular size and shape over a large size spectrum. One of the simplest methods is the regular droplet formation by the release of one liquid through a nozzle into a nonmiscible second liquid. This droplet generation can be performed very regular if constant streams of both liquids are generated. High symmetry in the flow arrangement and, therefore, high regularity in the droplet generation are achieved if both liquids are brought together in a coaxial arrangement. The release of these droplets and therefore their volume are determined by both the flow rates, the viscosity of the embedding liquid, the liquid–liquid interface tension, and the wetting behavior of the liquids on the surface of the nozzle.

The droplet size can further be steered by the geometry of the droplet generator. A high velocity, namely a high volume flow rate, of the embedding liquid leads to smaller diameters of the released droplets. At high flow rates, the instantaneous formation of droplets at the nozzle can be suppressed and a lamella or even a small beam or jet of the inner liquid is formed by the shear forces of the fast-flowing outer liquid. The effect of reduction of the diameter of the streaming inner liquid by a high flow rate of the outer liquid is called *flow focusing*. This small liquid jet relaxes at a certain distance from the nozzle by forming small droplets. Even this process can be performed very regularly, resulting in a narrow size distribution of the generated droplets.

Polymer particles can be generated by this microfluidic technique when a liquid monomer or monomer mixture is applied for the droplet-forming liquid, which is released by the nozzle. The monomer liquid should also contain an initiator for starting the polymerization. Mostly, thermal activation of radical initiators is used for starting a radical chain reaction for polymerization. But the addition of a thermal initiator to the monomer mixture makes the liquid sensitive for spontaneous polymerization, which can lead to blocking of the tubes, capillaries, and the nozzle. It is recommended to add the initiator only immediately before droplet generation in order to avoid the risk of early polymerization start. The best way is the addition of initiator solution by a micromixer in front of the droplet generator in order to minimize the residence time between mixing and droplet release. Alternatively, the initiator can be added to the droplet-embedding carrier solution. But in most cases, this type of initiation is not sufficient for complete polymerization of larger droplets.

5.6.2 Particle Formation by Photochemical Polymerization of Monomer Droplets

Much more convenient than the thermal initiation of polymerization is the application of a photochemical process. This strategy allows a safe decoupling between mixing of reaction components and starting of the droplet polymerization (Serra and Chang, 2008). The liquid handling and droplet formation can proceed in complete analogy to the above-described thermal procedures. But the monomer mixture with all components, including the photochemical initiator for polymerization, can be prepared before droplet generation without risk of early starting of the monomer reaction and blocking of tubes and devices. The only required measure is the exclusion of light with sufficient energy for the stimulation of the photoinitiator. Therefore, either the inlet capillaries and tubes microfluidic have to be built of dark wall material or the whole arrangement has to be kept in the dark. A radical-forming photochemical process with an initiation in the near-UV range is typical for this photochemical particle formation. In this case, it is sufficient to exclude the short-wavelength part of the visible spectrum, and it is possible to work under orange and red light.

The photochemical process is performed by a focused light beam after the droplet release. Thus, it is possible to transform the small liquid droplets very fast into a hardened particle. In the arrangement, care has to be taken that no or not much light can be scattered into the nozzle region in order to avoid undesired polymerization in the nozzle.

The monomer mixture is often completed by a certain content of a cross-linker for making particles stable and nonsensitive against polymer-dissolving solvents. A low degree of cross-linking can promote a certain swellability of the polymer matrix that supports the diffusive penetration of analyte molecules into the particle for sensing.

Compact polymerized and densely cross-linked microparticles are less suitable for sensing because the analyte molecules cannot diffuse into the particle matrix. Such particles can only carry reagent molecules or recognition sides on their surface.

5.6.3 Formation of Gel Particles by Photopolymerization of Monomer Solution Droplets

A high permeability of the particle matrix for analyte components can be achieved if the particle is formed by a gel instead of a compact polymer. Such microgel particles can also be generated by in situ photopolymerization of microfluidically prepared droplets. The only difference is that the mass polymerization of the monomer mixture is substituted by a solution-polymerization-like process inside solvent-containing droplets. Aqueous solutions of water-soluble monomers are used for the production of sensor particles for water-soluble analytes.

The monomer mixture is composed of an aqueous solution containing the water-soluble monomer as the main component, a smaller addition of a water-soluble cross-linker, the water-soluble photoninitiator, and molecular or NP additives for sensing. This mixture is mostly applied in the inner capillary of a coflow arrangement for serial generation of droplets. The droplets are formed by injection of this monomer mixture solution into the water-immiscible embedding outer liquid. The polymerization in these droplets is started by UV radiation shortly after the droplet release (Fig. 5.7). During the polymerization, polymer chains are formed that are partially cross-linked by the added cross-linking monomer. The concentration of the cross-linker determines the density of bridges between the macromolecular chains. As a result, a 3D polymer network is formed in which the solvent fills the pores between the polymer chains. Recognition molecules and NPs for sensing can be trapped inside this gel matrix, but smaller analyte molecules have a nonlimited access to the inner part of the sensor particle.

The only disadvantage of gel microparticles is their low mechanical stability in comparison to compact polymer particles. They have enhanced sensitivity against shear stress and have to be handled with certain care. But particles of this type can often be stabilized by drying and can easily be handled in the dried state. Normally, re-swelling to the gel state can be performed without problems. It is indicated, typically, by a strong increase of particle volume due to uptake of the solvent.

Bisacrylamide-cross-linked polyacrylamide is a convenient material for making sensor microparticles for applications in an aqueous environment. UV-sensitive asymmetric ketones are frequently used photochemical radical formers for polymerization initiation.

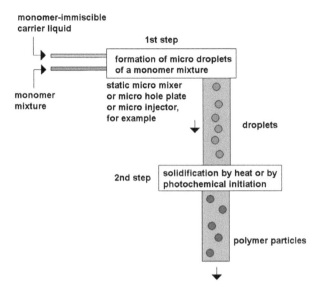

Figure 5.7 Formation of microgel particles by photopolymerization of droplets consisting of an aqueous solution of monomers in a microcontinuous flow process.

5.6.4   Particle Formation by Emulsion Polymerization

The generation of NPs by emulsion polymerization is a long-known and technically well-established process. It is applied for batch synthesis of polymer materials in mass production, too. The name is related to the start situation, where a low- or non-water-soluble monomer is distributed in droplets inside a water-based emulsion. The formation of NPs is due to a micelle-based mechanism. The trick is a combination of thermally induced polymerization with a phase transfer process between monomer droplets and the aqueous phase in which the NPs are formed. The transfer of the more or less hydrophobic monomer is promoted by enhanced temperature and by the presence of micelle-forming surfactants, which also stabilizes the emulsification of the monomer in the beginning of the process. Often, a water-soluble radical former is applied for the thermal initiation of the radical polymerization.

The radical polymerization starts in micelles and the growth of polymer chains leads to the formation of particles. Their growth is determined by the density of nuclei, on the one hand, and limited by the available monomer, on the other hand. After complete consumption of the monomer, a suspension of polymer particles is obtained. The particles are distributed inside the aqueous phase. The dispersed state is stabilized by the surfactant. Ionic surfactants cause an electrical charge of the polymer NPs that are responsible for an additional stabilization of suspension by electrostatic repulsion. A thermodynamically stable colloidal solution of NPs can be formed by this effect.

The typical size of NPs from emulsion polymerization is in the range between some tens and some hundreds of nanometers. In principle, the whole range from the mid-nanometer level up to about one micron can be addressed. It has to be kept in mind that an increase in volume between about 20 nm and 800 nm (diameter ratio of 40) is related to a volume factor of 64,000, which means that the volume of NPs from emulsion polymerization can really be varied to a large extent. A tuning of particle size is easily possible by the variation in surfactant concentration. Normally, the particle diameter decreases with increasing surfactant concentration. This effect is mainly due to the increase of micelle concentration and nuclei density in the start phase of polymerization. In addition, temperature, monomer type, emulsion quality, and monomer content influence the product particle size and homogeneity.

The shape of particles from emulsion polymerization is controlled by interface tension and micelles. Due to this fact, spherical particles are obtained, in general. Under special conditions, other shapes can be generated, for example, if an in situ assembling of growing NPs takes place before termination of the polymerization process. Under these conditions, significant yields of ellipsoidal, dumbbell-like, astragal-like, or branched polymer particles are formed (Visaveliya and Köhler, 2014).

For sending purposes, it is important to integrate recognition and transduction functions into the polymer NPs. The simplest way is the postpolymerization immobilization of additional molecules with special chemical functions on the particle surface. Particles from ionic emulsion polymerization can easily be surface-functionalized by electrostatic binding due to the high surface charge of the particles that is originated from the ionic surfactants. An emulsion polymerization in the presence of sodium dodecyl sulfate (SDS), for example, leads to negatively charged polymer NPs that can be easily connected with molecular cations, for example, molecules with quaternary ammonia groups. An emulsion polymerization in the presence of a cationic surfactant as cetyl trimethylammonium bromide (CTAB) leads to positively charged polymer particles that can be electrostatically bonded to anionic recognition molecules such as DNA, for example.

A more sophisticated way of particle functionalization is the integration of anchor or linker groups at the surface or inside the swellable polymer matrix. An introduction of such groups succeeds, for example, by copolymerization. Therefore the anchor or linker molecules have to be equipped with polymerizable functions as acrylate groups, for example. A special strategy is the in situ integration of dye molecules or of small metal NPs during the emulsion polymerization process. Such techniques are dependent on the solubility and phase behavior of the applied molecules and small inorganic NPs that have to be incorporated.

5.6.5 Particle Formation by Physical Precipitation

Physical precipitation is a simple route for the generation of micro- and nanoparticles. In principle, it is applicable for different types of material. In contrast to reactive particle formation, the chemical species of forming particles is not converted during the particle formation process but is already present in the dissolved state. The typical case is the precipitation from molecular solutions. This can concern polymer solutions as well as solutions of low-molecular-weight substances, organic as well as inorganic compounds.

Fast and homogeneous generation of nuclei is the essential precondition for any homogenous particle formation. Beside the start of the nucleation process, even the termination of nucleation should be synchronized in order to avoid later superposition of particle growth and proceeding formation of new nuclei. The time window for nucleation should be small in comparison with the time needed for the particle growth. A short nucleation phase in the first step is the basis for a homogeneous growth of particles in the second phase of particle formation.

In the growth phase, the most important demand is to keep the forming suspension in a stable state. Therefore, uncontrolled aggregation and sedimentation of particles and their adhesion on reactor walls have to be avoided. In the case of microparticles, the sedimentation and aggregation can be suppressed, mostly by moving the liquid and applying strong convections. In the case of NPs, the formation of thermodynamically stable colloidal solutions is the best way for a homogeneous particle growth phase and for obtaining a stable product.

In a wider sense, the formation of salt-like particles by precipitation can be regarded as physical particle generation. In the case of these particles, the particle-forming substances are present in solution in the form of ions. The particles are generated by the interaction of cations and anions under the formation of ion crystals. Particle formation is induced by a shift of the solubility product, typically. This can be achieved, for example, by addition of one of the two particle-forming ions, by cooling, or by an exchange of ligands of a change in ligand reactivity, for example, by protonation or de-protonation.

For polymer particles, the method is based on the segregation of solvent and dissolved macromolecules after a change of solvation conditions. The addition of a nonsolvent or a change in pH or in concentration of ions is a typical strategy for the initiation of such a precipitation process. Typically, coagulation of dissolved macromolecules is initiated at the beginning. But care has to be taken that complete aggregation of the coagulated polymer material does not take place. Homogeneous particle formation is supported by fast mixing of the polymer solution and the coagulation-inducing additive for a homogeneous start of precipitation, on the one hand, and intensive stirring or shaking for avoiding the undesired aggregation of particles, on the other hand.

For production of composed micro-/nanoparticles, the precipitation process can also be initiated in a material matrix. The mobility of NP-forming reactants is a precondition for this approach. One possibility is the in situ formation of inorganic NPs in a polymer matrix, which is solidified by polymerization, afterward. An alternative is the precipitation by interdiffusion of the solvent and precipitation components in a swollen particle matrix. A very high rate of particle generation is achieved by a combination of microdroplet formation and precipitation. The particles are composed of a sodium-sensitive ionophore and a chromo-ionophore embedded in a poly(vinyl chloride) (PVC) matrix of microparticles. The particles have a diameter of about 10 μm and are formed by precipitation with a rate of about 20 000/s (Tsagkatakis et al., 2001).

5.6.6 Emulsification by Static Micromixers

The formation of emulsions is the initial key step for particle generation by emulsion polymerization, but it is also important for forming NPs by direct solidification of monomer droplets by polymerization (*suspension polymerization*). Stirring and shaking are the mostly applied techniques for emulsion formation in conventional procedures in laboratories as well as in production facilities.

Static micromixers offer an interesting way for emulsification by a continuous-flow process. The formation of small droplets of a monomer mixture can be achieved by multilamination or related microfluidic techniques. Multilamination mixers, as well as split-and-recombine or superfocusing mixers, are typical microdevices for the generation of emulsions. In general, the homogeneity of droplets in emulsions from a microfluidic process is higher in comparison with conventional stirring. Typically, thin lamellas of both the immiscible liquids—the aqueous phase and the monomer mixture—are formed. These liquid lamellas are then spontaneously collapsing into small droplets of the monomer phase that are then distributed in the aqueous carrier phase.

The application of micromixers for emulsification demands for particle-free reactant liquids and a careful control of the initiation of polymerization. Uncontrolled early particle formation or even an early start of polymerization in liquids can cause an increase the liquid viscosity or other flow-related properties and can finally lead to serious disturbing of the process or complete blocking of the reactor.

5.6.7 Microhole Plate Emulsification

Microhole-plate devices are a special type of microreactors for continuous-flow emulsification. In contrast to micromixers, not only lamellas of liquids are formed, but the formation of two liquid phases is also directly coupled with the generation of droplets. Hole plate reactors contain a perforated plate with a 1D or 2D array of holes. The typical hole diameter is in the lower or mid-micrometer range.

The monomer mixture for particle generation is pressed through the hole array into the nonmiscible carrier phase (Fig. 5.8). Water, silicon oil, or a perfluorinated organic liquid is used, normally, as the carrier phase. A high frequency of droplet generation is achieved by a high number of holes in the hole plate and by high flow rates of the applied liquids.

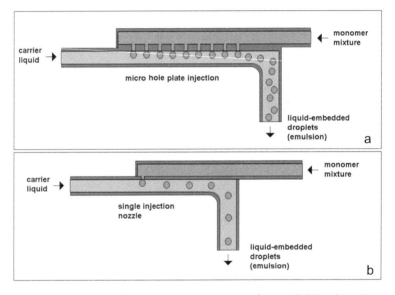

Figure 5.8 Example of a microhole arrangement for parallel droplet release for efficient generation of microparticles via an emulsification process of monomer mixtures.

The emulsification can be performed either by jetting of the monomer phase into the slower-moving carrier phase or by a cross-flow arrangement. In this latter case, the droplet release from the hole plate is controlled by the shear forces induced by the carrier-phase flow in the orthogonal direction to the monomer motion through the holes of the plate. A re-coalescence of released droplets can be suppressed by the application of a surfactant, which stabilizes the liquid–liquid interface.

In principle, microhole plate emulsification can supply a comparatively narrow size distribution of produced droplets, resulting in high homogeneity of produced particles. The size homogeneity is much better than in the case of emulsification by stirring or shaking, normally. But certain deviations in droplet size can be caused by pressure differences and velocity gradients of the streaming liquids. These effects lead to a certain variation in the conditions of droplet release from the single holes of the hole plate and finally to deviations in the droplet sizes.

5.6.8   Serial Droplet Release

The best way for the production of particles from droplets is the serial production of droplets under identical droplet release conditions. This serial production of droplets can be regarded as a special type of emulsification. Microfluidic droplet generators are very well suited for such emulsification processes. The low Reynolds numbers in microfluidic devices offer the best conditions for reproducible local droplet motions. Even in the case of spontaneous formation of local circular convections, the streaming conditions are reproducible, droplets can be generated in a constant frequency, and the size of released droplets can be kept constant.

A simple device for microfluidic droplet generation is a branched channel. Y- or T-shaped channel geometries have been applied for droplet generation and are usable for particle production. The droplet quality can be controlled by total volume flow rates and by the flow rate ratios of monomer mixture and carrier liquid. The robustness of droplet release process can be enhanced by the application of nozzle-like structures in the Y- or T-junctions.

The branched-structure microfluidic droplet generators include also an easy possibility of in situ mixing of components of monomer mixture. For this purpose, two or several inlet channels are arranged to bring different liquids directly to the droplet generator position. The different components are merged together immediately in front of the releasing nozzle, but the completion of mixing takes place after droplet release during the further droplet transport through the microchannel by a segmented-flow regime. This two-phase flow induces a regular circular convection inside the droplets, which can cause a fast mixing of components. The segments can further be relaxed into spherical droplets by enlarging the diameter of the microchannel or by release of the segmented flow into a larger reaction vessel. As an alternative, the polymerization and solidification can be initiated inside the flow segments. This leads to in-freezing of the segment geometry. As a result, rod-like particles are obtained. The length and aspect ratio of these particles are determined by the volume of the single fluid segments and the diameter of the segment-guiding microchannel.

An in situ mixing of reactants immediately before the droplet formation is of particular interest for the generation of polymer and polymer–composite sensor particles. It offers the possibility to bring together monomer and initiator, different monomers for in situ mixing, as well as the mixing of monomers and sensitive molecular, biomolecular, or NP components. This way, polymerization or other reactions between particle-forming components starting too early and provoking an increase in liquid viscosity, undesired separation effects, or other troubles in the material transport and droplet formation can be avoided.

Regular droplet generation can also be achieved by a coaxial fluidic arrangement between the carrier liquid and the liquid monomer mixture. In these coflow-droplet generators, the monomer liquid is injected through a capillary or nozzle, which is placed in the center of an outer tube or capillary transporting the carrier liquid. A co-centric parallel flow of carrier and monomer liquid is realized by this arrangement. In dependence of flow rates, capillary diameters and wetting conditions, droplets are formed in the dripping or in the jetting mode (Fig. 5.9). The quality of particles can even be improved and the particle diameter can be further reduced by the combination of the coflow droplet release with the flow focusing principle (Fig. 5.10) Particles are formed from monomer droplets by chemical solidification. Stable particles are advantageously obtained by light-induced photopolymerization. The photoinitiation can be achieved in a continuous process by conducting the serially formed droplets through the beam of focused light. Typical monomer mixtures consist on a base monomer, a photoinitiator, and a certain portion of a cross-linker in order to stabilize the polymer matrix and to control the mechanical properties and the swellability.

Microfluidics is very well suited for the generation of droplets with uniform size and shape (Serra et al., 2007). Shape and size can be tuned by the construction principles and the geometry of the droplet generators and by varying flow rates, flow rate ratios, viscosity, and interface tensions (Chang et al., 2009).

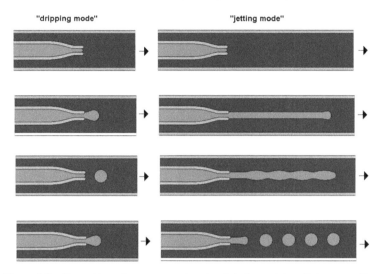

Figure 5.9   Formation of monomer microdroplets for serial particle generation in a coflow arrangement by dripping mode (left) and jetting mode (right).

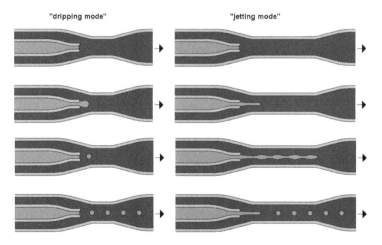

Figure 5.10   Combination of coflow and flow-focusing principle for formation of monomer microdroplets for serial particle generation in a coflow arrangement by dripping mode (left) and jetting mode (right).

The principle of coflow droplet generation is particularly interesting for the generation of regular sequences of droplets with diameters in the submillimeter and in the upper- and mid-micrometer range. In combination with in situ polymerization it is a convenient strategy for engineering different types of NPs, including double particles, microparticle necklaces, Janus particles, rod-like particles, composite particles, and core/shell particles (Chang et al., 2009). The integration of inorganic NPs like metal or oxide NPs is important for the microfluidic generation of composite microparticles for transducer applications (Chang et al., 2009).

Droplet formation can be performed either by a periodic droplet release at the nozzle (*dripping mode*) or by the formation of a liquid jet of streaming monomer in the center of the streaming carrier phase and subsequent relaxing of the liquid beam into a series of small droplets (*jetting mode*). The dripping mode is typical for lower flow rates, and the jetting for higher flow rates and higher shear forces. Both modes can supply very narrow size distributions of microdroplets.

The direct coupling of droplet generation with a subsequent photochemical initiation of a radical chain polymerization is a convenient method for converting the droplets into solidified particles. Complex compositions of microparticles can be realized by mixing of the desired components into the monomer cocktail. These components can be immobilized automatically inside the formed particle by the formation of the polymer network—either by encapsulation or by physical or direct chemical bonding to the forming polymer matrix.

Flow-focusing droplet generation is a special variant of the coaxial droplet generation. The flow-focusing effect is induced by a reduction of the inner diameter of the outer capillary in the coflow device at or shortly after the position of the nozzle. The enhancement of flow velocity of the carrier phase due to the reduced channel diameter causes an enhancement of shear forces and leads to a reduction of the droplet size in the dripping mode or to a reduction of the liquid beam diameter and, subsequently, a reduction of droplet and particle size, too, in the jetting mode.

5.6.9 Microfluidical Generation of Hierarchically Constructed Microparticles

The regular generation of droplets by release from a capillary or nozzle can be extended for the generation of composed particles by a combination of two or several coupled droplet release steps. In particular, nested or serial coflow arrangements are well suited for the generation of Janus-type, core/shell, multicore/shell, and core/multishell particles (Serra et al., 2013b; Wang et al., 2014).

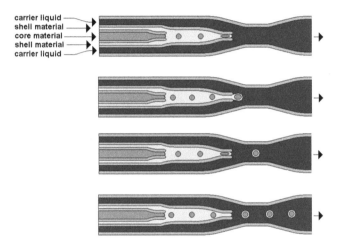

Figure 5.11 Coflow arrangement for microfluidic generation of core/shell microparticles.

A possibility of the generation of core/shell microparticles by the microfluidic coflow technique is shown in Fig. 5.11. First, a liquid core/shell structure is generated by the release from the core material from the central small capillary of a coaxial nested-capillary arrangement. This droplet is embedded in a nonmiscible phase of a second monomer or monomer mixture. This phase is forming a liquid shell around the core material when it is released from the second nozzle. This core/shell droplet forms by injection into the nonmiscible outside carrier phase. A typical material combination consists on a non-water-miscible monomer mixture in the core (e.g., alkylacrylates and bisacrylates), a water-soluble monomer for the shell (e.g., polyacrylamide), and non-water-miscible oil (e.g., silicon oil) as the carrier phase. The particles are subsequently formed by subsequent solidification of droplets, whereby photochemical initiation of radical polymerization is the most convenient way.

Figure 5.12 shows a scheme and a scanning electron microscopy (SEM) image of a microparticle carrying a mixture of metal-NP-incorporating polymer NPs on its surface. This particle is a typical example of a hierarchically constructed particle. Particles of this type are of interest for sensing as well as for catalytic applications.

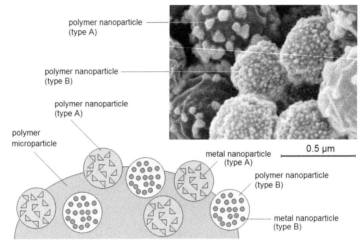

Figure 5.12 Scheme and SEM image of the surface of a hierarchically constructed microparticle carrying two types of polymer nanoparticles incorporating differently shaped small nanoparticles of two different metals (Li et al., 2017).

5.7 Special Sensor Microparticles for SERS Sensing

The basis for the application of silver nanoparticles (SNPs) in sensing is the so-called surface-enhanced Raman effect. The classical effect of Raman scattering was postulated by Smekal and later proved experimentally by Raman (Smekal–Raman effect). It is due to the resonant exchange of portions of energy between optical photons and molecular vibrational excitation states of the scattering molecules. This leads to a reduction of photon energy by the amount of this vibration energy (normal Raman lines) or to an enhancement of energy of scattered optical photons (anti-Raman lines), if energy of higher vibrational states is superposed to the energy of the scattered photon.

The Raman effect is very interesting for analytical purposes because it supplies information of molecular structure due to the characteristic vibrational resonance energies. In contrast to electron resonances—which are normally related to visible or UV light and supply broad and less characteristic lines—the vibrational lines have a much smaller bandwidth. They are characteristic for certain chemical structures and states because these resonances are directly connected with the strength of chemical bonds and the masses of vibrating atoms. The electromagnetic resonances of these vibrations are in the middle-infrared (MIR) range. This is the reason why a vibrational characterization is mainly related to IR spectroscopy. IR spectroscopy is one of the most fundamental techniques for chemical characterization and is applied in qualitative and quantitative analytics. A big disadvantage of IR analytics is the low IR transparency of a lot of materials. Measurements in an aqueous environment suffer particularly from high IR absorption by water, which makes IR investigations of biological samples difficult in many cases. Here, Raman spectroscopy is a very attractive alternative. Visible or near-infrared (NIR) light can be used instead of the longer-wavelength MIR range. In the short-wavelength range, much more materials and solvents, in particularly water, are transparent. In addition, very powerful light sources, optics, and detectors are available. Thus, Raman spectroscopy is a very attractive strategy for analytics in the optical range. The possibilities of spectral and geometrical separation of incoming and outcoming light and the possibility of focusing make Raman spectroscopy also useful for application in small volumes or even for characterization of microscopic objects like single cells. The characteristic patterns of Raman peaks are a very powerful tool for specific characterization. These can be used for single substances and simple mixtures as well as for the characterization of complex composed systems. Recently, the fingerprint methods in SERS sensing are far enough developed for the discrimination of single bacterial strains (Walter et al., 2011).

A significant disadvantage of classical Raman spectroscopy is the high ratio between the required illumination and the detectable scattered light. This leads to a comparatively low sensitivity of Raman spectroscopy, and therefore, high concentrations of analyte species have to be present in the sample. Thus, classical Raman spectroscopy is well suited for characterization of pure substances or mixtures with high concentrations of components as well as for reaction monitoring in chemical syntheses. But it is less suitable for low concentrations or trace analytics. This fact is a serious limit for the application of Raman spectroscopy for many bioanalytical problems, where the interesting molecules are not present in molar or millimolar but in micromolar or even lower concentration.

Therefore, the detection of SERS was a real breakthrough. Analyte molecules that adsorb on a metallic surface can cause Raman signals that are several orders of magnitude stronger than the classical Raman signals. Particular high SERS signals have been found on silver. Enforcement factors of a million and more have been observed by use of SNPs for Raman enhancement. This made the detection of molecules down to the nanomolar range—and in some cases—in still lower concentrations possible.

Reconsidering this background of SERS spectroscopy, it becomes clear that metallic silver-containing particles are very interesting for sensing purposes. The simplest way is the direct application of NPs in solution or in cell suspensions in order to measure SERS spectra of surface-adsorbed molecules. But in this strategy there is an intrinsic dilemma between the desired high concentration of SNPs in a suspension, on the one hand, and the effect of the analyte composition on the stability of the silver colloid and toxic effects of the high-concentrated silver colloid on biological objects in the analyte, on the other hand.

These problems can be significantly reduced by the application of SNP-containing micro- or nanocarrier particles. The idea of these particles results from the strategy of a partial decoupling between the molecules and other objects in the analyte and SNPs, which are essential for the SERS effect (Fig. 5.13). The connection between the SERS-active metal NPs and the carriers ensures a high local density of SNPs but avoids a high silver loading of the complete analyte solution.

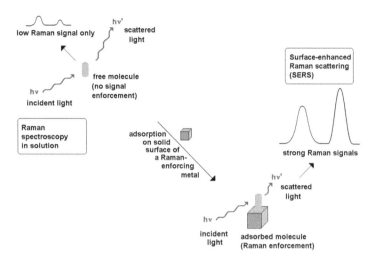

Figure 5.13 Surface-enhanced Raman scattering: strong enhancement of Raman signals by electronic interaction between analyte molecules and metal surfaces or metal NPs.

There are several preparation strategies for composite particles for SERS sensing. The simplest approach is the immobilization of SNPs at the surface of a larger carrier particle. The SNPs can be synthesized separately and secondarily coupled with the surface of the sensor particle, or they can even generated by in situ deposition on the carrier particle. Examples of metal NPs and composite sensor microparticles for SERS applications are shown in Table 5.3.

A special strategy is a two-step process with the deposition of small SNPs at medial or lower density at the surface of the carrier particle, followed by a metallic enforcement of the small immobilized particles by silver-catalyzed chemical silver deposition. Thus, a high but regular silver loading on particle surfaces can be achieved.

Organic as well as inorganic materials can be used for carrier particles. An easy way for obtaining carrier particles in high homogeneity is the synthesis of polymer particles by in situ photopolymerization of microfluidically generated droplets of a reactive monomer or prepolymer mixture. Monomer or prepolymer droplets in the range between ~100 and 500 μm can easily be generated by a flow-focusing or a coflow arrangement (Serra et al., 2013b). In this technique, droplets of the reaction mixture are formed by the interface tension after the release of the reaction mixture into an immiscible carrier phase. These droplets pass a photochemical reaction module, in which the photopolymerization is initiated by the irradiation of the droplets by focused UV light. Depending on the composition of the reaction mixture, solid or gel-like carrier particles with typical diameters in the submillimeter range are obtained. SNPs can then adsorbed or chemically bond at the surface of these polymer particles. As a result, SNP-covered polymer sensor particles are obtained.

Table 5.3   Examples of SERS sensor particles.

Material	Size	Measurements/ Analyte (examples)	Ref.
Au/Ag core shell	30–110 nm (tunable)	Thionaphthole	Samal et al., 2013
Polyacrylamid/Silver composite particles	0.3 mm	Adenine	Köhler et al., 2013
Polyacrylamid/Silver composite particles	0.1 mm	Histidine, adenine, serial flow measurements	Visaveliya et al., 2015b
Polyacrylamid/Silver composite particles	0.4 mm	Lactic acid, pantothenic acid	Visaveliya et al., 2015a
PMMA-based polymer/metal nanoassemblies	0.15–5 µm	Adenine, cytosine, tryptophane	Visaveliya et al., 2017
Silver nanotriangles, chitosan-coated	65 nm edge length	SERS imaging of cells, cell proliferation	Potara et al., 2013

The microfluidic technique can also be applied for the deposition of SNPs inside the gel matrix of a larger carrier particle if the monomer mixture for droplet generation contains a polar solvent beside the monomer and prepolymer. In this case, SNPs with hydrophilic surface can be colloidally dispersed in the monomer mixture and are in situ–embedded inside the polymer matrix during the photopolymerization of the fluidically formed monomer/prepolymer droplets. Even these SNPs can be chemically enforced if the gel matrix is permeable enough for silver ions and a reducing agent. This silver enforcement can be performed after the microparticle synthesis and washing in an aqueous suspension, for example, by simultaneous or alternating rinsing with a silver salt solution and a solution of the reducing agent.

A third way is the direct in situ generation of SNPs inside the gel matrix of the photopolymerizing microparticle (Köhler et al., 2013). Therefore, the silver precursor is directly mixed with the monomer/prepolymer cocktail. An in situ mixing of the metal precursor solution with the monomer mixture is recommended in order to avoid an undesired early chemical initiation of polymerization by radical formation due to starting redox reaction of silver ions. The reducing agent for the silver ions should be contained in the monomer mixture in order to avoid an early start of the SNP formation and possible coagulation and precipitation. An overview on formation of SERS–sensor microparticles in the form of silver/polymer composites is shown in Fig. 5.14.

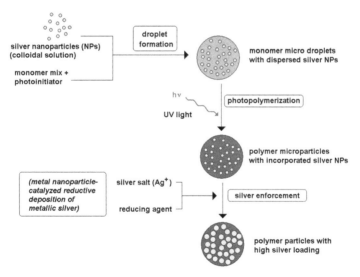

Figure 5.14 Formation of silver-enforced polymer composite particles for Raman sensing.

5.8 Particle Encoding

Beside the transduction functions, micro- and nanoparticles are often used for labeling of samples, droplets, or larger particles. In many cases, it is of interest to get the possibility of individual recognition of single particles. This can be achieved by specific encoding of each particle. An optically readable encoding is the most convenient way if the particles are large enough for getting the required optical resolution. Typical strategies are the use of distribution patterns of fluorescence dyes, writing up of microbarcodes by focused laser beams, or distribution of an ensemble of fluorescent NPs inside a microparticle (Meldal and Christensen, 2010).

Different strategies for combinatorial coding for particle labels are shown in Fig. 5.15. On the one hand, the linear principle of the classical barcode can be transferred to microparticles by strips or by the incorporation of smaller particles in a linear order. On the other hand, coding in spherical microparticles can be performed by incorporating smaller particles of different shape, color, or size or a combination of different features.

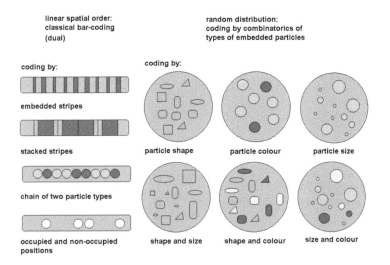

Figure 5.15 Composed microparticles for labeling: barcoding-like linear principles (left) and combinatorial labeling by embedding ensembles of smaller microparticles in the matrix of a larger microparticle (right).

5.9   Special Surface Architectures for Particle-Based Sensing

Surface-immobilized particles are not to be counted in the "mobile spy concept" in a closer sense. But there are some sophisticated developments using NPs in surface-related environments that might become interesting for the development of sensor particles in the future.

Xiong et al. (2016) connected two plasmonic nanobeads by a molecular spring. The contraction of this spring caused a decrease of the particle distance, resulting in electronic coupling and a bathochromic shift of plasmon resonance, which means long-wavelength absorption. The expansion of the molecular spring led to an electronic decoupling of both the plasmonic NPs and was visible by short-wavelength electronic resonance. This measurement of mechanic forces can be coupled with analyte-dependent biomolecular motions. The authors showed that the coupling of the molecular spring with integrin allows to read out the concentration of reductive oxygen species (ROS).

Chapter 6

Application of Sensor Microparticles as "Mobile Spies" in a Technical Environment and in Living Systems

6.1 Sensor Particles at Surfaces and Interfaces

A convenient method for the detection of molecular interactions is the change of optical properties of a surface caused by the attachment of metal nanoparticles (NPs). This effect is used in the change of transmittance and in the reflectivity of smooth transparent surfaces by the specific reaction with nanobead-conjugated molecules. The strategy of modification of optical surface properties by particle binding is applied in the detection of complementary DNA oligonucleotides by binding to gold nanoparticles (GNPs; Csáki et al., 2001). The bead-supported interaction of complementary DNA is a typical example for highly specific biomolecular interactions (Fritzsche, 2001).

6.2 Sensor Particles in Microfluidic Compartments

Despite the fact that the optical method can provide a lot of data from chemical or biochemical systems, the specificity of molecular

Mobile Microspies: Particles for Sensing and Communication
Michael Köhler
Copyright © 2019 Pan Stanford Publishing Pte. Ltd.
ISBN 978-981-4800-14-3 (Hardcover), 978-0-429-44856-0 (eBook)
www.panstanford.com

detection is very low in many cases. This is partially due to the low specificity of the applied method in general: Absorption and fluorescence spectroscopy, photometry, and fluorometry supply more or less unspecific data if they are not coupled with a selective recognition principle, in addition. To another part, the low specificity is due to the special molecular system: All DNA molecules, for example, are constructed by the same molecular building units, and it is rarely possible to distinguish different nucleic acid sequences by optical spectroscopy. Molecules with the exact same brutto composition can drastically differ in the essential biomolecular information, which is coded in the order of the building units. In such cases, a direct contact between a primary transducer for molecular recognition cannot be avoided. But the application of contact sensors is difficult to realize in experimental, diagnostic, analyses, or handling systems using microfluidic compartments as microcapsules, microdroplets, or microfluid segments. In these systems, "mobile microspies" are absolutely required for the readout of molecular or biomolecular data.

An important advantage of the microbead-based strategy of primary transduction it is universal applicability. Thus, sensor microparticles can be used for measurements of simple chemical parameters as pH, inorganic ions, and dissolved gases, as well as for more complex organic compounds and complex constructed biomolecules in an analog way.

The determination of pH during the cultivation of microorganisms and other cells is a central task for cultivation monitoring. Metabolic activity is connected with change in pH, normally. The strength of pH shift is dependent on cell density, on the character and intensity of metabolism, and on the available substrates and gas exchange. Therefore, pH monitoring is of interest both for the biotechnical production in large fermenters and for experimental or screening processes in very small volumes.

Funfak et al. (2009) used pH-sensitive fluorescent microparticles for the monitoring of bacterial growth in microfluidic compartments. The particles with a diameter of about 3 μm responded with a change in fluorescence intensity on a shift in pH. This effect was due to the pH dependence of fluorescence quantum yield of the immobilized dye. A bacterial growth causes a shift in the pH in the cultivation liquid in the volume of about 0.5 μL and was indicated by a shift in the fluorescence intensity. The method is also suitable to monitor the transition from respiration to fermentation (digestion) by a strong decrease of pH that is due to the acidic metabolic products. The shift of pH and the subsequently changed fluorescence signal from the microsensor particles can also be used for the determination of toxic effects and critical concentrations of toxic substances in microtoxicological screenings. Highly resolved dose–response functions could be determined by stepwise variation of concentrations of effectors in series of microfluid segments with single volumes in the submicroliter range. The response of microorganisms on the different effector concentrations could be read out by monitoring the fluorescence intensity of sensor particles inside the microfluid segments, which report on the intensity of their metabolic activity.

Optical detection is crucial for noncontact signal transfer in chemical sensing. Spectral discrimination allows multiple sensing strategies, which can also be applied for particle-based chemical sensing (Nagl and Wolfbeis, 2007). Beside pH, oxygen concentration is a second key parameter in cell cultivation, and new strategies and probes for measuring these parameters at the millifluidic and microscale are required (Demuth et al., 2016). Oxygen microsensor particles are interesting for a larger spectrum of applications, among them miniaturized cell cultivation in microdroplets and in microfluidic systems (S. Sun et al., 2015). Instead of fluorescent particles, sensor particles incorporating phosphorescence dyes are applied for oxygen sensing. They respond on oxygen by a lowering of their luminescence quantum yield caused by the triplet character of the two-atomic oxygen molecule. The oxygen sensor microparticles can be dispersed, too, in microfluid compartments and report about the metabolism of microorganisms, on the oxygen consumption in the cultivation medium by a growing cell population (Horka et al., 2016). In analogy to the pH monitoring, they can also be used for the determination of dose–response functions in microtoxicological investigations (Cao et al., 2015). Particles with suitable combinations of incorporated sensor dyes can also be used for simultaneous determination of different analysts, for example, oxygen and pH (Ehgartner et al., 2016). The particle-based oxygen sensing is not restricted to the visible range of electromagnetic spectrum. Scheucher et al. (2015) showed that even the infrared

(IR) region can be used for luminescent particles for miniaturized oxygen measurements.

6.3 Sensor Particles in Tissues and Cells

6.3.1 Fluorescence-Based Cell and Tissue Characterization

The staining of cells and tissues by dyes is a long-used technique. On the one hand, the contrast of the microscopic structure is enhanced. On the other hand, different structures can be made visible by application of dyes with different chemical properties due to a preferential binding and accumulation of dye in special parts of cells, for example, in cell organelles, cytoplasm, or membranes. The differentiation of binding sites by the chemical affinity of staining dyes represents a certain kind of reporting about local chemical properties. This strategy of selective binding for getting local information by microscopic imaging can be transferred to NP labeling, too (Deng and Wang, 2014). Such particle labels are, in general, more stable than molecular tags (Bai et al., 2015). High sensitivity was achieved by staining with fluorescent inorganic NPs as quantum dots. But these particles can be corroded and release toxic ions into cells, which affect the cells strongly and disturb a monitoring of cellular processes (Resch-Genger et al., 2008).

Organic NPs, for example, polymer NPs, with incorporated organic dyes represent a convenient alternative to the toxic inorganic tags (Chen et al., 2016). But their fluorescence intensity is lower, in general, due to the limited number of fluorophores inside the particles. The fluorescence intensity from dye molecules in such NPs is particularly limited by self-quenching effects that increase with increasing density of dye molecules. Recently, it was found that this dilemma can be overcome by a type of dyes that show an enhancement of fluorescence in contrast to quenching in the case of high dye concentration and molecular aggregation. Such an enhancement of quantum yields was found for tetraphenylethene-based dyes and can be applied in organic nanocapsules with high dye concentration (S. Xu et al., 2016). These particles show high biocompatibility. The immobilization of specifically binding peptides leads to selective markers for tumor cells, for example.

Nanoscaled fluorescent particles with special chemical sensitivity attract particular interest due to the possibility to get spatially resolved information from biological materials. In addition, they can combine the readout of intracellular concentration with high temporal resolution because the typical response times can be reduced to the millisecond scale (Lu and Rosenzweig, 2000).

The pH value is a simple but, nevertheless, a very important parameter. It can be determined locally in tissues or inside single cells by introduction of optically readable sensor particles. This strategy was followed by Kreft et al. (2007). They prepared polymer microcapsules containing a pH-sensitive dye. Zhang et al. (2014) applied a similar strategy by using fluorescein isothiocyanate (FITC)-labeled microcapsules, which had been introduced into HepG2 and endothelial cells for readout of cell-internal pH by measuring the fluorescence intensity from single sensor particles. A similar approach was used for pH measurements on a dendritic cell line (JAWS II) from mouse bones by Alexa Fluor (AF488) dyes embedded in a poly(diisopropylamino)ethylmethacrylate (PDPA) capsule matrix (Liang et al., 2014). Instead of organic material, inorganic NPs can be used for cell-internal pH sensing, too. Chu et al. (2016) modified silicon NPs by pH-sensitive molecules in order to realize pH-sensitive fluorescent NPs, which could be applied for measuring pH in HeLa and MCF-7 cells. Fluorescent metal ion sensor particles are frequently also sensitive for pH changes.

An example for metal ion imaging is given by quantum dot (QD)-based fluorescence resonance energy transfer (FRET) particles for calcium sensing (Zamaleeva et al., 2015). This system is excellently suited for calcium concentration imaging in the submicromolar and nanomolar ranges. CdSe/ZnS QDs act as energy donors, dye-conjugated chelate ligands for calcium ions (1,2-bis(o-aminophenoxy)ethane-N,N,N',N'-tetraacetic acid [BAPTA]), and acceptors for excitation energy.

Yu et al. (2013) used fluorescent carbon nanodots for the visualization of cells in dependence on the H_2S content. Therefore, they coupled naphthalimidazide with carbon NPs. In the absence of H_2S, only the original C-dot emission was observed. After reduction of the azide groups to amino groups, energy transfer from the absorbing carbon particles to the naphthalin units (FRET) takes place and could be detected by a longer-wavelength emission. In result, different H_2S-dependent long-wavelength emission from HeLa and L929 cells was measured and visualized in the microscopic images.

NPs of amphiphilic block copolymers, including fluorescence dyes, had been developed for the measurement of hypochloric acid inside lysosomes (Wang et al., 2017). Therefore, monomers had been connected with both the donor and the acceptor dye. The particles were formed by a precipitation process and had a typical size of 30 nm. They had applied for measuring the concentration of hypochloric acid inside lysosomes of Hela cells.

The determination of copper concentrations in the brain tissue of rats was succeeded by the application of NPs, which consisted of lanthanide coordination polymers. The copper ions occupy the coordinative binding sites of included sulfosalicylic acid that causes a quenching of the NP fluorescence (Huang et al., 2015).

Table 6.1   Applications of sensor particles in cell-related chemical sensing

Analyte/ Measurement	Particle type	Biological sample/target	Ref.
Al^{3+}	Carbazole-conjugated block-copolymer NPs	Vero cells	H. Liu et al., 2013
ClO^-	Iridium-complex-modified silica NPs	RAW 264.7 cells	Zhang 2015
Cu^{2+}	BODIPY-functionalized silica NPs	SMMC-7721 cells	Yu et al., 2014
Cu^{2+}	Fluorophore-labeled polyacrylnitril NPs	SK-BR-3 cells	Lee et al., 2014
Cu^{2+}	Lanthanide coordination polymer NPs	Rat brain	Huang et al., 2015
Erythrosine	Fluorescent organic NPs	Food control	Mahajan et al., 2016
H_2S	Naphthalimidazide-modified carbon dots	HeLa and L929 cells	Yu et al., 2013
Heparin	APTES-functionalized Si NPs	Blood serum	Su-dai 2016

Analyte/ Measurement	Particle type	Biological sample/target	Ref.
HOCl	Polymer FRET particles	Lysosomes in Hela cells	Wang et al., 2017
NO	Gold-NP-labeled silica nanocapsules	3T3 cells	P. Rivera-Gil, 2013
O_2	Ru-complex-functionalized silica NPs	Rat C6 glioma cells	Xu et al., 2001
O_2	Pt-complex-functionalized silica NPs	Rat C6 glioma cells	Koo et al., 2004
O_2	LBL-coated polymer NPs	Developed for intracellular analysis	Guice et al., 2005
O_2	Phosphorescence dye-doped polymer NPs	Bacterial growth in microfluid segments	Cao et al., 2015
O_2	Phosphorescent NPs for endocytosis	Fibroblasts, neuron cells, cortical cells, and others	Dmitriev et al., 2015a
O_2	Pt-complex-doped polymer NPs	Mouse liver tissue	Dmitriev et al., 2015b
O_2	Phosphorescence dye-doped PSPVP NPs	*E. coli* cultures in microdroplets	Horka et al., 2016
pH	SNARF-1-labeled polymer microcapsules	NRK fibroblasts, MDA-MB435 breast cancer cells	Kreft et al., 2007
pH	Gold/ poly(vinylalcohol)-polyacetal core/shell NPs	CHO cells	Stanca et al., 2010
pH	FITC-labeled microcapsules	HepG2 and endothelial cells	Zhang et al., 2014
pH	AF488-labeled PDPA microcapsules	JAWS II cells (mouse bone)	Liang et al., 2014

(Continued)

Table 6.1 *(Continued)*

Analyte/ Measurement	Particle type	Biological sample/target	Ref.
pH	N-doped carbon NPs	T24 cells	Shi et al., 2016
pH	Molecular-modified silicon NPs	HeLa and MCF-7 cells	Chu et al., 2016
SDS, SDBS	Cyclodextrine-conjugated fluorescing NPs	KB cells	Xu et al., 2012
Thiols	Molecular-functionalized polymer NPs	Biological thiol compounds	Ang et al., 2014
Thiols	Fluorophore-coupled gold NPs	Thiol compounds in aqueous solution	J. Xu et al., 2016

NPs that respond to small drug molecules can be applied for the monitoring of drug release and drug targeting. Such NPs are usable for imaging of drug distribution in tissues (Feng et al., 2010). Free radicals could be detected by the interaction of two conjugated polymers in fluorescent NPs (J. Wang et al., 2011).

Fluorescent NPs can be used for measuring gene activity. The gene expression level can be monitored by imaging of the primary products, the messenger RNA (mRNA). Polymer particles with a cross-linked shell containing DNA had been applied for the detection of mRNA in mouse macrophages (Z. H. Wang et al., 2013). Rapid detection of micro-RNA (miRNA) was achieved by nanoassemblies of single-stranded DNA conjugated with silver nanoclusters. After hybridization the miRNA could be detected by fluorescence from the nucleic acid/nanocluster adducts (Yang and Vosch, 2011).

Enzymatic activities can be characterized by the connection of a test substrate with signal generation. For this purpose, microparticles can be applied as supporting materials for the sensing process. These particles can be regarded as transducers so far as they are responsible for the first signal conversion. An example of this particle type and sensing process was reported by Frisk et al. (2008). Their task was to develop a sensor for botolinus toxin, one of the most poisonous substances. This substance is able to mediate the cleavage of peptide bonds. This property is used in the bead-supported detection. Therefore, a fluorophore-connected peptide was immobilized on the particle surface. In the presence of the toxin, the peptide is cleaved and the fluorophore is deliberated from the particle surface and can be proved optically.

The caspase-3 activity was measured by the conjugation of core/shell QDs (CdSe/ZnS) with the fluorescent protein mCherry (Boenemann et al., 2012). The semiconductor NP acts as a primary absorbing material and exciton donor in a FRET process. The fluorescent protein acts as the energy acceptor and supplies the luminescence that can be detected and used for microscopic imaging.

Particles can also be used for the characterization of other enzymatic acitivities. Chemburu et al. (2008) applied surface-modified polymer-coated silica microspheres for the characterization of a phospholipase function. Therefore the particles had been functionalized by an anionic lipid that quenches the fluorescence of the polymer shell of the microparticles. In the presence of the phospholipase the quencher is deliberated and the fluorescence intensity of the polymer matrix increases.

Fluorescent NPs are well suited for imaging and for selective labeling of tumor cells (Feng et al., 2014). A special detection of cancer cells was realized by the use of fluorescent NPs with a molecular imprinted shell. The particles have been prepared by a precipitation method and imprinting with sialic acid (Liu et al., 2017). Fluorescent NPs can also be generated by direct incorporation of fluorophores into a copolymer matrix of NPs by using fluorescent monomer components. A FRET system for intracellular imaging of cancer cells was constructed by use of a polycaprolactone block copolymers with incorporated dyes (Kulkarni and Jayakannan, 2017). A FRET system including a photoswitchable spiropyran was developed for specific labeling of human prostate cancer cells.

Therapeutic and monitor functions can be combined by the application of dye-conjugated cancer-targeted GNPs. By use of these particles a complete transduction chain, including the therapeutic effect, is constructed (W.-H. Chen et al., 2014): This transduction chain uses the fact that during the apoptosis of the cancer cells an enzyme (caspase-3) is produced that cleaves specifically the peptide sequence Asp-Glu-Val-Asp (DEVD). A peptide including this amino acid sequence is used for conjugating GNPs with a fluorescence dye. The fluorescence of the dye is completely quenched by the very efficient energy transfer between the excited fluorophore and the metal due to the short distance. The application of GNPs on the cancer tissue induces the apoptosis of the cancer cells (therapeutic effect). The apoptosis is accompanied by the production and release of caspase-3. This enzyme cleaves the peptide linker between the dye and the NP. In result, the fluorophores are deliberated from the particle surface and there fluorescence is recovered by the reduction of energy transfer due to the increasing distance between the energy donor and the acceptor. The increase in fluorescence can be observed by fluorescence microscopy. This allows spatially resolved monitoring of the therapeutic effect of the GNPs. The dye-conjugated NPs act as therapeutics and nanosensors for monitoring at the same time.

Table 6.2 Application of sensor particles for the detection or monitoring of biomolecules and cells

Biological target	Particle type	Transduction principle	Ref.
Amino acids	Calixaren-functionalized Au and Ag NPs	Fluorescence	Makwana et al., 2017
Bacteria cells	Antibody-functionalized silica NPs	Binding and fluorescence	X. Wang et al., 2004
Bacteria cells	Carbon NPs with aptamer-conjugated QDs	Fluorescence switch-on	Duan et al., 2015
Bacteria cells	Au NPs coupled with bacteria-specific aptamers	FRET quenching	Jin et al., 2017
Cancer cells	Antibody-functionalized polymer NPs	Binding and fluorescence	J. Liu et al., 2013
Cancer cells	Aptamer-functionalized dipeptide NPs	Fluorescence	Z. Fan et al., 2016

Biological target	Particle type	Transduction principle	Ref.
Cancer cells	FRET-conjugated polymer NPs	Fluorescence/ FRET	Kulkarni and Jayakannan, 2017
Cancer cells	Immunofunctionalized polymer NPs	Photoswitching	M.-Q. Zhu et al., 2011
Cancer cell, apoptosis monitoring	Au NPs with peptide-linked fluorophores	Caspase-3-induced fluorescence recovering	W.-H. Chen et al., 2014
Cancer cells vs. glycan affinity	Sialic acid–imprinted polymer NPs	Binding and fluorescence	Liu et al., 2017
Cancer cell (prostate)	Dye-conjugated fluorescent polymer NPs	Fluorescence/ FRET, photoswitching	Rad et al., 2016
Caspase-3 activity	mCherry-conjugated CdSe/ZnS NPs	Fluorescence/ FRET	Boenemann et al., 2012
Cholesterole	Curcumin-integrated SAL on silica NPs	Fluorescence	Chebl at al. 2017
Colon carcinoma cells	Antibody-conjugated QDs on magnetic beads	Fluorescence imaging	Ahmed et al., 2013
DNAse	Ru(II)-labeled gold NPs	Fluorescence quenching	Cao et al., 2017
Dopamine	Fluorescence dye-conjugated polymer NPs	Imaging of fluorescence quenching	Qian et al., 2015
Dopamine	DNA-coupled copper NPs	Fluorescence quenching	Wang et al., 2015
Dopamine	Au NPs in combination with graphene QDs	Recovering of QD fluorescence by GNP coagulation	Lin et al., 2016
E. coli ORN178	Glycan-functionalized silica NPs	Binding and fluorescence	X. Wang et al., 2011

(Continued)

Table 6.2 (*Continued*)

Biological target	Particle type	Transduction principle	Ref.
Epithelial cell adhesion molecule	PEGylated graphene QDs and MoS$_2$ nanosheets	Fluorescence switch-on	Shi et al., 2017
Fibroblasts and macrophages	Dye-doped polystyrene NPs	NIR fluorescence lifetime measurement	Hoffmann et al., 2013
Furin activity in cells and tumors	Self-quenching NIR fluorophore NPs	Fluorescence switch-on	Yuan et al., 2015
Glutathione	Manganese dioxide–modified NPs	Luminescence	Li et al., 2014
HeLa cell imaging	Ho^{3+}-doped NaYbF$_4$ up-converting NPs	Up-conversion emission	P. Du et al., 2017
Human serum albumin	Self-quenching NIR fluorophore NPs	Fluorescence switch-on	X. Fan et al., 2016
Listeria	Antibody-functionalized silica NPs	Binding and fluorescence	Z. Wang et al., 2010
Lymphocytes	Antibody-functionalized silica NPs	Binding and fluorescence	Lian et al., 2004
Lsyosomes	Polythiophene NPs	Two-photon fluorescence	S. Zhao et al., (2017)
mRNA (gene expression)	Knedle-like core/shell NPs	Fluorescence/FRET	Z. H. Wang et al., 2013
mi-RNA	Polythymine-linked copper NPs, 3 nm	Fluorescence, after primer extension	Chi et al., 2017
Sialylation of cell surface	Cy3-/Cy5-coupled silver NPs	SNP-enhanced FRET	T. B. Zhao et al., 2017
Tumor cells	Iridium-doped mesoporous silica NPs	Phosphorescence, bioimaging	Tu et al., 2017

Biological target	Particle type	Transduction principle	Ref.
Tumor cells	Aptamer-conjugated magnetic particles	SERS imaging	C.L. Sun et al., 2015
Tumor tissue	Nd^{3+}-doped LaF_3 NPs	Fluorescence-based temperature measurement	Carrasco et al., 2015
Tumor tissue	Copper sulfide NPs, ~20 nm diameter	Photoacoustic imaging	Yang et al., 2014

The high penetration ability of polythiophene NPs was applied for the targeting of lysosomes and for the visualization of intracellular structures and deep-tissue imaging by two-photon excited fluorescence measurements (S. Zhao et al., 2017). Lipid-modified fluorescent NPs had been introduced for addressing of cells and labeling (Feng et al., 2010). Such particles are suitable for enhancement of fluorescence in membranes, membrane stacks, and lipid layer–rich structures as the Golgi apparatus. High contrasts for the imaging of cell membranes were achieved by red-emitting fluorescent polymer nanoparticles (FPNs) that could be applied for two-photon excitation (Liu et al., 2015).

6.3.2 Phosphorescence-Based Cell and Tissue Characterization

The quenching of the phosphorescence of dyes, forming triplet states after electronic excitation, can be used for convenient measurement of oxygen concentration. An imaging of oxygen distribution can be achieved if small phosphorescent particles are introduced into the investigated objects (see Section 3.2.2). The mapping of NP phosphorescence intensity can be used for characterization of oxygen distribution in living cells and tissues (Dimitriev and Papkovsky, 2012).

6.3.3 Cell and Tissue Characterization by SPIONs

Besides optical transduction, electrical, electrochemical, photoelectrochemical, and magnetic properties of NPs can be used for biosensor applications (Willner and Willner, 2002). The abbreviation "SPION" stands for *superparamagnetic iron oxide nanoparticle*. These magnetic nanoprobes can be trapped and manipulated by external magnetic fields. In combination with immobilized recognition molecules and other molecular binding sites, they are very interesting for localized sensing (Rahman et al., 2015).

Ahmed et al. (2013) developed composed fluorescent magnetic particles. Therefore, they functionalized the surface of iron oxide particles by cetyl trimethylammonium bromide (CTAB) in order to charge the particles positively. In a further step, negatively charged QDs had been immobilized by electrostatic forces on the surface of the magnetic particles (Fig. 6.1).

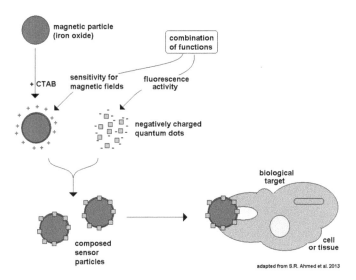

Figure 6.1 Formation of fluorescent composed magnetic particles by CTAB-mediated electrostatic immobilization of quantum dots on the surface of iron oxide particles (Ahmed et al., 2013).

The colloidal state of SPIONs has to be stabilized by chemical surface functionalization. Thus, spontaneous aggregation can be prevented. At the same time, the surface functionalization can be used for integration of special recognition molecules as well as for the loading with biological active molecules as medical drugs. The surface character of SPIONs plays a crucial role for their biocompatibility, their distribution in a body, for specific targeting and labeling, and for their biokinetics (Gupta and Gupta, 2005). The interaction of magnetic NPs with biomolecules, among them proteins, can be controlled by suitable surface functionalization and macromolecular coating (Sakulkhu et al., 2015).

Hydrophobic particles for magnetic contrast enhancement can be realized by incorporation of SPIONs in the hydrophobic matrix. Therefore, Floris et al. (2011) developed SPION@liposomes hybrid nanoarchitectures. Liver tissue–specific labeling was achieved by bovine serum albumin (BSA)-coated SPION assemblies. The authors (Zhang et al., 2012) synthesized nanoclusters of SPIONs with a shell of BSA molecules with a diameter of about 50 nm by assistance of ultrasound.

The special peptide CREKA was conjugated with SPIONs in order to get sensor NPs with high affinity to fibrin (Song et al., 2014). These particles are accumulated in the case of microthrombosis. Thus, it was possible to detect a small thrombus in rats optically and by magnetic resonance imaging. SPIONs conjugated with Alzheimer-specific antibodies could be used for the labeling of Alzheimer tissue in mice models (Solberg et al., 2014).

Functionalized magnetic NPs are of particular interest for the characterization of cancer tissue and for targeted therapy (Widder et al., 1980). They can combine the analytical and sensor function with therapeutic activity. The SPIONs show high biocompatibility and can be applied for the binding of anticancer drugs (Polyak and Friedman, 2009) or for local hypothermal cancer treatment.

The cell-specific binding of SPIONs in cancer tissue is very useful for the contrast enhancement in magnetic imaging (Cunningham et al., 2005). Many cancer cells show an overexpression of folic acid receptors (FARs) that can be used for specific cancer cell labeling by folic acid–functionalized NPs. Alpsoy et al. (2017) activated the SPIONs by tetraethylorthosilicate (TEOS) and coupled folic acid by using 3-amino-propyltriethoxysilane (APTES) as a linker molecule.

A special contrast enhancement effect for the magnetic resonance imaging of lymph nodes was achieved by using mannose as a recognition structure. Muthiah et al. (2013) developed SPIONs by integrating mannan-coated SPIONs in mannose-functionalized poly(ethylene glycol) (PEG) NPs. Gastric cancer cells had been visualized specifically by application of SPION-containing sensor particles with specific binding to cancer-related small interfering RNA molecules (Y. Chen et al., 2013).

Table 6.3 Examples of SPIONs for biomedical applications

Application (small-sized NPs)	Size	Surface functionalization	Ref.
Magnetic contrast enhancement	8 nm	Polydehydroalanine	Von der Lühe et al., 2015
Cancer cell labeling for magnetic resonance imaging	14 nm	Folic acid	Alpsoy et al., 2017
Cancer cell labeling for magnetic resonance imaging	4 nm	Hyaluronic acid	Thomas et al., 2017
Cancer analysis and therapy (HeLa, MCF-7, U87, A549, L929)	14 nm	Carboxylated luteoline	Alpsoy et al., 2017
Cancer theranostics by SPION-supported gene delivery	4 nm	Hyaluronic acid	Thomas et al., 2017
Application (composed NPs)	**NP construction**	**Chemical function**	**Ref.**
Magnetic contrast enhancement	SPIONs in liposome NPs	Hydrophobic	Floris et al., 2011
Magnetic contrast enhancement	SPIONs in Fe-polymer microparticles	Carboxyl groups	Borges et al., 2015

Application (small-sized NPs)	Size	Surface functionalization	Ref.
Liver-specific magnetic contrast enhancement	SPION-BSA nanoclusters	BSA	Zhang et al., 2012
Labeling for combined MRT and PET	Fe_3O_4/silica core/shell NPs	Cu^{2+} loading	Patel et al., 2010
Labeling for combined MRT and fluorescence	SPIONs and fluorophores in silica NPs	Nanoporous structure	Gogoi and Deb, 2014
Stem cell labeling for magnetic contrast enhancement	SPION-dye and peptide assemblies	Peptide	Lee et al., 2009
Cancer cell labeling for magnetic resonance imaging	SPIONs in albumin NPs	Folic acid	Ma et al., 2015
Magnetic contrast enhancement and therapy (HepG2 cells)	SPIONs in polymer NPs	Doxorubicin release	Zhu et al., 2017
Magnetic contrast enhancement for lymph node imaging	SPIONs in polyethyleneglycole-NPs	Mannose	Muthiah et al., 2013

MRT, Magnetic resonance tomography; PET, positron emission tomography.

Thomas et al. (2017) developed SPIONs functionalized with hyaluronic acid for theranostic applications. They used the fact that several frequent cancer types such as head, neck, and colon cancers are marked by an overexpression of the CD44+ receptor that is specifically binding on hyaluronic acid (Zhou et al., 2000; Toole, 2004). The used SPIONs had a diameter of 4 nm only and had been integrated into micelles of poly(L-lysine) and hyaluronic acid. The incorporation of small SPIONs into larger NPs for contrast enhancement in magnetic resonance imaging was also applied for FAR-overexpressing cancer cells. Therefore Ma et al. (2015) entrapped SPIONs from polyol synthesis in albumin nanospheres and conjugated folic acid on the surface of these composed particles.

The strategy of combination of labeling for magnetic imaging with drug application by SPIONs could be extended to a light-controlled drug release: Zhu et al. (2017) reported about nanoassemblies constructed by photosensitive diblock copolymers embedding SPIONs and drugs at the same time. They could demonstrate that the photostimulated release of doxorubicin causes an increased cytotoxicity toward HepG2 cells. Borges et al. (2015) incorporated SPIONs inside microparticles of a catechol-based Fe^{3+} coordination polymer for magnetic image contrast enhancement.

Cu^{2+}-labeled NPs had been developed in order to realize local labeling suitable for positron emission tomography (PET) and magnetic contrast enhancement at the same time (Patel et al., 2010). The particles consist on a SPION core and a copper ion–modified silica shell. A dual signal function was also realized by combining the magnetic contrast enhancement by SPIONs with particle fluorescence. Therefore, larger silica NPs encapsulating SPIONs together with fluorophores have been prepared (Gogoi and Deb, 2014).

6.3.4 SERS-Based in situ Characterization

A chemical characterization of cytoplasm by surface-enhanced Raman spectroscopy (SERS)-active NPs demands for suitable sensing particles and a strategy to bring these particles into the cells. The in situ SERS measurement of NO in 3T3 cells succeeded by the application of small amino-functionalized GNPs incorporated in a silica nanocapsule (Rivera-Gil et al., 2013). The construction principle and the formation of such amino-functionalized nanocapsules is illustrated in Fig. 6.2.

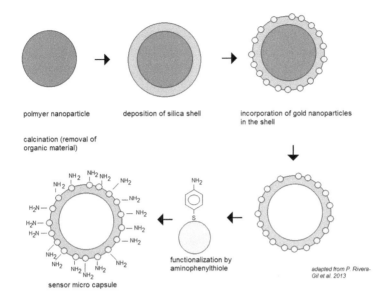

polmyer nanoparticle   deposition of silica shell   incorporation of gold nanoparticles in the shell

calcination (removal of organic material)

sensor micro capsule

functionalization by aminophenylthiole

adapted from P. Rivera-Gil et al. 2013

Figure 6.2 Construction principle and formation of amino-functionalized silica nanocapsules (Rivera-Gil et al., 2013).

Highly resolved SERS spectra have been obtained by the application of surface-modified silver nanotriangles. The particles with an edge length of 65 nm had been modified by *para*-aminophenol and chitosan. They are suitable for incorporation into A549 cells and can be used for generation of SERS images at different Raman wavelengths (Potara et al., 2013).

6.3.5   FRET-Based in situ Characterization by FRET-Active Particles

In analogy to fluorescence, the resonant energy transfer between an absorbing dye and a fluorescing molecule (FRET) can be used for a local labeling of tissues and cells with high sensitivity. One strategy in bead-based FRET sensing uses the energy transfer between an exciton (energy) donor connected with the analyte and an exciton (energy) acceptor on the particle surface. A sufficient high specificity of the method can be achieved by chemical recognition caused by specific binding of the analyte molecule with specific binding sites on the particle surface. The excitation energy is only transferred to the luminescing molecule if this binding event between the labeled analyte and the particle surface occurs. After several binding events, the luminescence intensity from the particles is strongly enhanced. The particles work as detecting elements and as concentrators for contrast-enhancement by enrichment of the light-absorbing molecules at the particle surface.

In the case of a nonexcitable analyte, FRET labeling is required. Therefore, analyte molecules have to be linked to the label at first. The binding on the particle is then the second step in the process chain for sensing. Alternatively, the FRET labeling can also occur after the binding of the analyte on the particle surface. In this case a second binding process on the particle surface is required that has to be specific for sites on which analyte molecules are attached. The use of a FRET-labeled complementary binding antibody could be suitable strategy for such a labeling. The simplest way for FRET-based sensing is applicable if the analyte itself acts either as an energy donor or as an energy acceptor. In this case, the functionalization of the particle surface by the complementary FRET partner is sufficient for the energy transfer and an additional conjugation of analyte or of the analyte binding complex on the particle surface can be avoided.

Chapter 7

Living Models: Particle-Based Sensing and Communication in Nature

7.1 Particles and Biological Big Data

The biosphere has developed several efficient principles and tools for efficient communication. For judging these principles as models for technical sensing and communication systems, it has to be kept in mind that the development of biological systems is always subordinated under the Darwinian principle of maximization of the number and fitness of offspring. In contrast, the performance of technical sensing and communication should be judged independently from a specific population and their individual properties and requirements.

Despite this difference, information storage and processing in living nature and the use of particles, therefore, supply interesting models for technical sensor systems. At first, living nature had created substance bond storage and processing of information. In particular, the molecular information storage system of nucleic acids is very efficient. It had been very successful in biological evolution, and this success is evident by the universal use of this principle in all living beings. It is both applied for intraorganismic communication as well as communication between organisms of the same species and between very different species. Horizontal gen (genetic

Mobile Microspies: Particles for Sensing and Communication
Michael Köhler
Copyright © 2019 Pan Stanford Publishing Pte. Ltd.
ISBN 978-981-4800-14-3 (Hardcover), 978-0-429-44856-0 (eBook)
www.panstanford.com

information) transfer by plasmids is a typical example of this type of biological communication.

A further very important principle is the packaging of information sets into particle-like objects. The large chromosomes of higher organisms are a typical example of packed information-carrying molecules. These particles are formed by assembling giant DNA molecules with construction proteins into condensed supermolecular objects. These objects can be further manipulated—for example, in the process of cell division or by the unification of haploid chromosomal sets by the formation of a zygote by fusion of male and female gametes.

On the more macroscopic scale, plant seeds and polls are particles for dissemination of information. It is interesting to look at the similarity and differences of the function and transport strategies for seeds and polls from the point of view of information transfer. Seeds contain the complete genetic information of a plant and are additionally equipped with a package of nutrients for the start phase for the development of a new individual. They are spread by different biological data mechanisms using very different transport ways, for example, by wind, water, birds, or other animals, which are distributing the seeds randomly in a closer or farther environment.

A particular class of information-carrying particles is represented by viruses. A virus can be described as a nanocapsule containing biological information coded by a DNA molecule. Viruses are particles with a diameter in the middle or upper nanometer range. They are composed of a certain set of biomacromolecules. And they are marked by a very regular construction. The arrangement of molecules and atoms is so regular that viruses can be crystallized and their structure can be recognized from X-ray diffraction. This is only possible because all viruses consist of the same kind and number of biomolecules and each atom in a single virus particle is found at exactly the same position as in all other viruses. This high regularity is achieved besides the fact that viruses can be constructed of thousands of molecules and can have molecular weights of hundreds of million Daltons.

Beside the storage of molecular and biological data in viruses, the complex body of so many molecules is formed by a self-assembling process. The formation of the virus shell and their core is controlled by molecular recognition functions. Each molecule finds its right place in the molecular assembly automatically by its own morphology and functionality. The structure of the whole virus particle is coded by a set of complementarities of the involved protein molecules. In addition to this astonishing coded molecular assembling, the shell of a virus also includes the function of recognition of its hosts, the ability to attack them specifically, and the deliberation of the virus DNA into the cytoplasm of the host cell for initiating the replication and transcription of the virus DNA into the virus proteins for multiplying the virus by use of the biomolecular machinery of the host cell. The fact that this multiplicity of tasks can be organized in a nanoscaled particle gives hope that technical particles are thinkable that also are able to respond to different tasks and can be formed and transported using mechanisms such as surface recognition and self-assembling.

7.2 The Hierarchy of Particle-Based Organismic Communication

Beside interorganismic communication, particles also play an important role in the transport of information in biological intraorganismic communication. The lowest structural level of information transport is given by small molecules as hormones, which are very important for the regulation of central functions of an organism. The next level is formed by macromolecules as proteins and protein assemblies. From the view of particle-based signal transport, even supermolecular assemblies as antibodies can be regarded already as nanoparticles (NPs).

The further subcellular level of intraorganismic communication is formed of exosomes. These objects are vesicle-like structures, but they are without the essential components of a living cell and are smaller and less functional than a real cell organelle. They are generated by exocytosis, which means that their membrane is originated from the cytoplasmic membrane of the exosome-releasing cell.

Finally, complete cells are released from tissues and contribute to signal transfer and distribution of important information over the whole organism. The generation and multiplication of cells of the immune system is a typical example of such a cell-based system of information transport. In this case, the cells combine the carrying of essential information with specific physiological functions as the recognition of viruses, bacteria or other strange and dangerous objects and the ability to bond them, to mark them, and to destroy them by endocytosis and metabolization.

The particle-based intraorganismic communication can reliably work only if the production of particles is always complemented by a continuous disappearing of particles. A healthy particle-based communication system needs a robust steady state of particle formation and particle degradation. The total density of particles has to kept nearly constant by equilibrated production and degradation mechanisms. As a result, each type of disseminated particle has a certain mean lifetime.

7.3 Communication by Exosomes and Other Vesicle-Like Compartments

Exosomes can be vary in size from the nanometer up to the micrometer range. They are transporting cytoplasmic fractions of their mother cell and molecules of specific recognition and transport functions that are embedded in their membrane.

The biological tissues are continuously releasing exosomes. The intensity of exosome production is dependent on the state of the tissues and their stress exposition. Thus, the density and character of circulating exosomes is always a mirror of the state of certain tissues. The distribution of exosomes by blood circulation and other body liquids over the whole organism can be understood as a multidirectional communication system (Fig. 7.1). It keeps all part of a body informed about the status of other parts, similar to the reports from events and in different parts of a country in a daily journal.

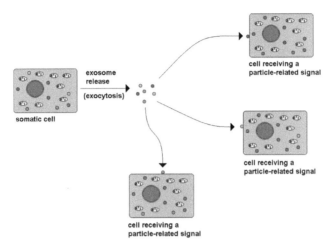

Figure 7.1   Multidirectional interorganismic communication by releasing and receiving of exosome-coupled molecular information by different cells, tissues, and organs.

Chapter 8

Visions: Particles in Future Technical Communication Systems

8.1   Information Distribution by Particles

The role of particles in communication has a lot to do with the directionality of information exchange. The development of human and technical communication systems is strongly related to the directionality of information exchange, too. A few examples can illustrate this fact: Oral and written letters have, in general, a sender and one individual or a small group of receivers. The spoken or hand-written message is the typical case for unidirectional communication. Talks and the exchange of letters can be understood as bidirectional communication.

The situation changed by the invention of book printing. Now, messages have been transferred from one emitter to many receivers. Book printing is the classical case for polydirectional information transfer. In principle, several more recent developed techniques for information transfer follow the directionality of book printing, for example, the printing and distribution of journals, broadcast, and television.

A principal new quality of information exchange was created with the birth of the World Wide Web. Now, an omnidirectional transfer of information has been realized, because every participant can

Mobile Microspies: Particles for Sensing and Communication
Michael Köhler
Copyright © 2019 Pan Stanford Publishing Pte. Ltd.
ISBN 978-981-4800-14-3 (Hardcover), 978-0-429-44856-0 (eBook)
www.panstanford.com

address messages to many or all other participants of the Internet communication. The Internet can be used as an electronic version of the old uni- or bidirectional communication, for example, by email or in a kind of polydirectional communication as in the form of group emails or WhatsApp messages, for example. But social networks and message systems such as Twitter or YouTube allow to deliver information from a lot of sources to a huge number of receivers simultaneously. In this type of communication, the bidirectionality of communication has lost its importance. The dominant mechanisms of information exchange are the sending-only mode, for example, by delivering a tweet, or the read-only mode, for example, by surfing through many tweets from different senders. This style of communication by send-only and read-only is very efficient as it is long known from journals, broadcast, and television (Fig. 8.1). And this style of communication matches perfectly the transport of information by particles.

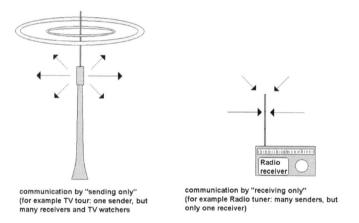

communication by "sending only" (for example TV tour: one sender, but many receivers and TV watchers

communication by "receiving only" (for example Radio tuner: many senders, but only one receiver)

Figure 8.1 Classical example of the polydirectional working sending-only/receiving-only principle.

Particle-carried information demands for writing systems, which allows to code information into a body, into particles, into a molecule, or into a molecular mixture, which means into matter, on the one hand. On the other hand, it demands for reading systems, which are able to transfer the information coded in particles coming from outside into an internal, system-compatible language.

The new favored principle of sensing-only in the Internet is a very old strategy in living nature. It is really what living beings are doing since billions of years. And these living beings are doing this by using matter, particles, molecules, eggs, and seeds. The encoded information is multiplied and distributed in the environment— sometimes over a large area (Fig. 8.2). The persistence and further tradition of information are a question of its matching to the environmental conditions; it's a question of fitness, adaptability, a question of competition and a certain portion of luck. And it has a lot to do with the timescales of transport, receiving, and exploitation of the transferred information.

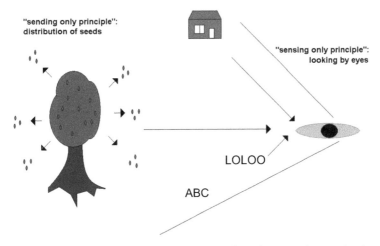

Figure 8.2 An old natural sending-only process based on particles: production and distribution of seeds from plants.

It is interesting to have a look at the potential of recent electronic data transfer systems. The internal electronic time requirements are small: Electronic switching can proceed in the picosecond range (10^{-12}–10^{-10} s). The microsecond range (10^{-6}–10^{-4}) gives a good rough orientation for the time need of the transfer of small messages and other data sets. Reading machines can operate these data in the millisecond range (about 10^{-3}–0.1 s), typically. Finally, the timescale of human perception, discussion, and the formation of meanings, intensions, and consequences for practical decisions are in the time span of minutes and hours, at least.

Molecular processes are slow in comparison to electronic processes in solid-state devices, in general. Fast chemical reactions for rearranging of atomic groups and for the forming and splitting of bonds are in the microsecond range, corresponding to the time need for rotation and local transport of atom groups. Thus, they are about a million times slower than electronic operations. But they have two advantages that are essential for the development of populations of organisms and might play an increasing role in the development of communication systems in future: (i) they are operated in three spatial dimensions, and (ii) they can be stored, in principle, over a very long time. The number of released information-containing particles of natural sending-only systems can be high: Some fishes can produce thousands of eggs, while a plant can produce ten or hundred thousands of seeds and millions of pollen particles. One colony of microorganisms can release billions of spores into the environment.

The huge potential of matter-coded information transfer becomes clear by a look at the orders of magnitude at the volume scale: 1 L of a substance can be used for making 1000 particles of 1 mL, 1 million particles of a volume of 1 μL, or 1 billion particles of a volume of 1 nL. These last-mentioned particles have a size of about 0.1 mm and are, therefore, larger than most biological cells.

Particles of the size of an average body cell are in the order of magnitude of 10 μm, corresponding to a volume of a picoliter (10^{-12} L), which means that a trillion of them can be made from 1 L of substance. The smallest bacteria have a volume of about 0.01 fL (10^{-17} L), which means that 10^{17} of them can be made from a 1 L portion of substance. Keep in mind that each of these cells has a genome of about 100,000 bytes as information content. Then, 1 L of these cells has an information storage capacity of 10^{22} bytes, a number that is equivalent to a stack of 10 billion of 1 TB storage. Imagine that for a height of 1 cm each, this memory stack would have a height of 100,000 km, which is about 8 times the diameter of our planet!

The principle achievable storage density of molecular storages is in the range of about 100 bytes/nm³, if small substituents—small atom groups—can be used for coding. This ultimate number would allow to storing 1 TB in 10^{10} nm³, which corresponds to a particle size of about 3 μm, approximately the size of a small cell. Remember that about 1 mL of matter is sufficient for producing 10^{13} particles with this information content.

These simple estimations illustrate the principle power of the particle-based sending-only mode. But it can be only applied if fast-working writing and reading systems are available.

How could the technical development in universal particle-based communication systems be imagined?

A central precondition for long-term development of such a strategy is the compatibility with the ecological systems on earth. Sustainability is a challenge with the same importance for communication systems as for energy production and material use. This means that the release of any kind of matter into the environment must not damage any ecosystem. The information-carrying particles should have a limited lifetime. They should be metabolized by ubiquitous organisms, for example, by soil bacteria. The components of particles should be re-introduced into the cycles of biological use without any residue.

These ecological challenges concerning the substance nature of disseminated communication particles suggest that organic substances have to be preferably used for information storage and transfer via micro- and nanoparticles. From the point of view of material recycling, biopolymers would be best suited, and the first look might be directed to peptides and DNA. Nature shows that these substance classes are well suited for information storage, and a second enormous advantage of these both biopolymer types is the availability of molecular tools for biochemical processing. But the use of these both biopolymer types for technical information distribution has also some drawbacks. Under active biological conditions—living soil, for example—the lifetime of these substances might be too short because of the always present biochemical destroyers, proteases for the peptides, and restriction enzymes for the nucleic acids. In addition, both substance classes have a high risk potential due to their central function in biological systems: they can be toxic or alter the molecular biological processes in organisms.

Therefore, other substance classes might be better suited for molecular information storage and transfer. Polysaccharides are also metabolized by many organisms. But there are not so many aggressive degrading molecules for carbohydrates released from organisms in the environment than in the case of proteases and nucleases. In general, polymers, which are the product of a condensation process, seem to be best suited for information storage by the order of molecular building units. They have the advantage of a systematic combination of strong and stable chemical bonds inside the elementary building units, for example, nonpolar covalent bonds, and less stable hydrolysis-sensitive bonds, which connect the elementary building units. All three mentioned classes of biopolymers—peptides, nucleic acids, and carbohydrates—match these requirements. But in principle, there are many other classes of condensation polymers that also meet these needs of structure and chemical behaviors. Their suitability depends on the serial steps in the communication chain:

1. The methods of transferring information into these molecules
2. The packaging of molecules into transferrable and storable particles
3. The distribution paths of these particles
4. The recognition and estimation of information-carrying particles by potential receivers
5. The depackaging of particles
6. The possibilities of readout of the molecular-encoded information by the receiving systems

Even the size of particles can be important for the ecological compatibility of distributed particles. Particles smaller than 50 μm represent a potential danger for human beings if they are distributed in the air, because they can be inhaled and damage the lungs. For such small particles it is particularly important that they be formed by materials that can be metabolized by animals and humans.

Despite the information transfer by single particles, the use of swarm strategies is under discussion for future sensing and communication procedures. Recently, such strategies are under discussion for lab applications (Lam et al., 2017), but they might become interesting for ex-lab communication in the future, too.

8.2 Writing Systems for Particle-Based Communication

Living nature had created different writing systems for molecular information. The template-controlled synthesis of nucleic acids during DNA replication for cell division, the transcription of DNA into RNA during the realization of the genetic code, and the translation of RNA sequences into proteins are very impressive mechanisms for the synthesis of information-carrying molecules. In particular, for nucleic acids, we have learned to transfer information from electronic devices into molecular sequences, too. Therefore, For this purpose, DNA synthesizers—chemofluidic automata—are used, which translate the digital electronic data into the sequence of bases in the synthesized DNA molecules. In contrast to the DNA synthesizers of living beings, the length of synthetized molecules is restricted to several thousand building units, the nucleotides, in a synthesis program. Nature can do it for very long molecules consisting up to millions of building units. And there is a second important difference: The technical synthesizers are macroscopic devices in the dimension of cubic meters, whereas the intracellular information transfer machinery has a volume of less than an attolitre (10^{-21} m^3).

Chromosomes are an interesting model for synthesis and packing of molecular information. The giant DNA molecule of a chromosome is generated by the molecular replication machinery. But finally, a compact object is formed instead of a long, thin, and sensitive thread. A typical chromosome of higher multicellular organisms is built by a DNA chain of about 10^8 nucleotides, which corresponds to a length of about 0.3 m. This extreme long molecule is packed by supermolecular mechanisms into an object of a few microns (10^{-6} m) by assisting proteins (histones) in a molecular self-organizing process. We learn from these base mechanisms of cellular DNA reproduction that such a combination of synthesis and packing of molecules into microparticles is realistic. Future technical particles for communication should have analog properties:

- Synthesis by microscaled machineries

- Packing of information-carrying molecules by supermolecular self-organization
- Forming of stable microparticles
- Packaging into a particle, which supports the recognition and depackaging by the receivers

For the last point, separate transducer functions have to be implemented. Probably, the initiation of estimation and uptaking of the information of a communication particle must be supported by several transduction channels. This would enhance the probability of recognition and use of the transported information. Therefore, optical imaging, electrostatic surface charging, fluorescence labeling, biocatalytic functions, or special chemical recognition of surface elements could be used simultaneously.

These challenges call for hierarchically constructed particles and a hierarchically organized information management system. There has to be an inner sphere mass store using the high capacity of molecular data storage, on the one hand. On the other hand, this core memory has to be capsuled by a shell that doesn't need to carry masses of information but should be able to present—or to send actively—the right data to the environment in order to draw the attention of interesting receivers to the particle. The particle-releasing system has to fabricate the particles complementary to the recognition and reading capabilities of the expected receiving systems.

In a certain way, viruses are a good model for such information-transferring particles. They consist of a core unit containing the virus DNA and a protecting shell, which is able to recognize cells that can be infected. But there is an important difference between the function of viruses and the task of communication particles. Viruses multiply their own information using the resources of the infected cells and destroy them, finally. The communication particles should multiply their information by generating a win-win situation with the information-receiving systems. They should neither destroy the receivers, nor should the information be simply multiplied. The goal of the particle-based information transfer should assist reading, processing, and merging of the information from the particle with the information pool of the receiver.

8.3   Reading Systems for Particle-Based Communication

Any receiving system has to be equipped with facilities or subsystems for finding particles carrying interesting information, for contacting these particles, and for extracting the data sets and translate these data into a system-compatible form. These challenges are an important reason for a minimum of complexity of any receiver system.

Two fundamentally different strategies can be chosen for the first step, the finding of interesting particles. One possibility is the filtering: This strategy needs a permanent stream of material carrying particles of interest. These particles can be dispersed and transported by air, water, or other liquid or gases. A filter-based search system must actively organize this material stream and needs; in addition, there should be an internal system for the picking, evaluation, and sorting of particles in order to select such particles, which are really of interest.

Probably, such systems will work with a nested system of filtering and sorting steps. In the first step, all potentially interesting objects will be separated. Set the case that message-carrying particles are forming 1 ppm of all mobile material; 1 g of material has to be processed in total for obtaining 1 μg of message particles in this first step. This step has to be fast but cannot be very selective. For judging the possible efficiency of such a step, it has to be taken in mind that a particle of a diameter in the order of magnitude of 10 μm corresponds to a mass in the order of magnitude of 1 ng. This means that about 1000 particles are selected from 1 g or about 1 million from 1 kg.

The really relevant particles have then been selected by a second sorting step. This process can even be time critical if particle-internal information has to be gained by a complex molecular reading process. This second selection can be accelerated considerably if each particle of potential interest is carrying an easy-to-read code that can be immediately recognized by the selection system. It is self-explaining that there have to be developed conventions for this fast selection procedure.

The selection of interesting particles from the environment by recognition in distance and directed picking is an alternative strategy. This strategy is only imaginable with a highly sophisticated, preferably noncontact optical detection system in combination with a high-performance micropicking system. The selection of interesting particles has to be subdivided into several steps demanding for specialist device features.

The receiving system should permanently screen its environment for potentially interesting particles. For this purpose a combination of imaging and scanning is required. Set a resolution of about 0.1 mm for the identification of interesting particles; an imaging system with a capability of 25 megapixels (5000 × 5000 pixels) corresponds to an area of ¼ m². For this resolution, the receiving system has to scan about 1200 images for overlooking an environment of a radius of 10 m (about 300 m²). This number of images increases up to 30,000 images if a resolution of 20 µm should be addressed. This number of images has to be taken and processed in 20 s in the case of a velocity of the moving receiving system of 1 m/s only. A resolution of 20 µm makes sense, if not only the existence of an information-carrying particle should be detected, but also particles should additionally be characterized from each other by specific optically distinguishable marks, for example, microbar codes.

If an interesting particle is found, the system has to move to it and pick it up. This device can be imagined as a sucking capillary like a pipette or like the fingers of a robot's hand. Finally, the selected and uptaken particle have to brought to an internal screening system that sorts the particles by more detailed surface information for deciding over the complete reading of internal particle information or releasing the particles as waste back into the environment.

A big advantage of this distance screening system in comparison with the filter system is that no permanent uptake of material from the environment is required. That should make this optical screening system much faster and more flexible. But its functions and construction are much more complex. The efficiency of this distance-oriented recognition system is strongly dependent on a complementary construction of a particle and imaging system. The information-carrying particles have easily to be recognized by the imaging system. They should give a strong optical contrast from the background materials, for example, by special structures, textures, and colors that can easily be distinguished by optical imaging. The electronic filtering of images for these particle features replaces the mechanical filtering of the other particle recognition strategy.

Optical imaging demands for particles of a certain minimum size. Despite the fact that even in a particle with submicrometer size a lot of information can be stored in molecular structures, such small particles can hardly be recognized and distinguished by scanning an environment by optical images. Thus, larger particles—say in the millimeter or submillimeter range—are preferred for optical imaging and the selective picking system.

It could be imagined that future particle-based communication systems are based on a dual strategy: For universal information screening, on the one hand, they will have a filtering system for easily accessible small particles as in dust or dispersed in water droplets in air, in soil pores, or everywhere. The filter-uptaken particles are small sized and are only roughly categorized without reading the internally stored information. In addition, they could be equipped by the optical-steered particle-picking system, which addresses larger and specifically labeled particles containing much more specific and much larger information sets.

Looking to the number and density of communication particles that could be released into the environment, it is worth to make a comparison with recent natural systems. Small communication particles with a volume in the order of magnitude of about 1 fL are comparable with bacteria in size. Typically, between 10^6 and 10^{10} living bacteria cells are found in 1 mL of natural soils. These values are corresponding with a total volume fraction between about 1 ppm and 1%. One billion of 1 fL communication particles (diameter of about 1 µm) would mean a mass fraction of 0.1% of the total mass of soil. Set a homogeneous distribution of such particles in the 0.1% density; 10 mL of soil is sufficient for storing an information set of every human being on the globe by an individual particle.

A similar calculation can be made for the distribution of particles in the air. The natural density of dust particles of about 1 µm diameter in unpolluted air is in the order of magnitude of 10,000/L (10/mL). This value is related to clean air. Dirty air has much higher particle concentrations. A particle recognition system has to filter about 40 m^3 air per hour for screening and sorting about 10^{10} particles, which is also the order of magnitude of the global human population. The global atmosphere has a huge capacity for particle storage and transport. About 10^{15} of these small particles can be stored in 1 km^3 at the above-mentioned particle concentration of 10/mL only. The particles are distributed and transported randomly. The communication is mainly based on the big number of these particles. The production and dissemination of these particles by sending systems can be compared with the **r** strategy in biological reproduction. It is highly **r**eproductive, **r**apid, and **r**andomly working.

On the other hand, the communication by larger particles is based on competence, high capacity, and intelligent cooperation. This part of the particle communication system is more related to the so-called **k** strategy in reproducing living systems. It is organized by concentration of much more information inside one single particle. Each particle is equipped with recognition labels. The dissemination, distribution, and transport of these particles are less randomly and target-related managed. The release system has the properties and interests of potential receivers in mind, already. In the environment, these particles are much rarer than the small communication particles. They are more used for information transport in special communication channels or by the use of special particle depot systems.

The typical size of these particles could be in submillimeter or millimeter range. If the small communication particles are compared with small bacteria by size, the competence-oriented larger communication particles correspond rather to small worms or insects. This analogy with living beings is also continued in the structure and internal organization of these particles. They are complex-constructed by subsystems that support fast identification and picking of particles by the receiver system and an efficient readout of stored information.

8.4 Hierarchically Structured Communication Beads

8.4.1 Particle Architecture and Storage Capacity

It is assumed that neither large nor small communication particles are uniform in their composition and structure. Both consist of different materials and molecules in order to achieve a maximum of information storage capacity and an optimal support for particle identification and information extraction by the receiver. These requirements always cause a hierarchical internal structure for the particles. This structure has to be generated by the particle-producing system.

A small particle of about 1 μm in diameter could have a weight of about 10^{-12} g or about 10^{13} Daltons. This means that this particle can contain about 10 million information-carrying macromolecules with a molecular weight of about 1 million Dalton that is in the order of magnitude of larger single protein molecules. It could easily be imagined that the formation of some packages of molecules would be very helpful for the reading system to decide which molecules have to be read out in detail and which portions of information are less useful for the receiver system and can be omitted. A hierarchical organization of the content of a particle is necessary for efficient access to relevant information. Similar to digital data files in the computer technique, each information-carrying molecule will be equipped with a header that says to the readout system quickly what type of information is stored in the following molecular sequence. In a similar way, each package of hundreds or thousands of molecules carries also such a header, which is able to give short and clear information to the receiver about what kind of information can be expected from the whole package of information molecules. And the small particles as a whole, which contain hundreds or thousands of these molecular packages, will also carry a header with a kind of content list for all included packages inside. This makes clear that already the small information particles should have a hierarchical structure if they should work properly and should be read out by addressed receivers efficiently.

The demand for a hierarchical internal organization is still stronger for the generation and structure of large communication particles. The much more precise targeting of these particles, the higher costs of their production, and their lower availability by occasional events make the intelligent hierarchical organization of these particles very important. Besides their size and information storage capacity, the higher production costs and the lower general availability of these particles demand for higher competence in bringing the relevant information to the most important receivers as fast and as completely as possible.

Looking to a possible internal structure of the larger communication particles, it is necessary to estimate their large storage capacity if a molecular storage system is assumed. At first it has to take a look at the principle size-related storage capacity. A particle of 1 mm diameter has a weight in the order of magnitude of 1 mg or 10^{19} Daltons. Set the case that a molecular bit needs about 20 Daltons for being coded, the maximum principle total storage capacity is in the order of magnitude of 0.5×10^{18} bits or—if 80% of the particle volume is required for organization and only 20% of the volume is available for storage itself—10^{17} bits.

A 1 mm particle could be filled with about a million of 1 µm particles. It is clear that efficient management of this large number of internal information packages requires further levels of a hierarchical organization. The spatial architecture of such hierarchically organized particles has to be adapted to the requirements of packaging and unpackaging of the sublevels and information carriers and on the requirements of fast writing and reading of the headers in the different levels of organization.

8.4.2 Behind the Borderlines between Materials, Devices, and Information Storage

The above-developed vision of a particle-based future communication system was drawn from the traditional view in which information is stored by a material and devices are constructed by materials. The stability of information as well as the function and reliability of devices are based on the stability of materials. This is the scheme of subdivision of the world into hardware and software. Devices and storages are forming the unchangeable hardware, whereas information represents the software.

In reality it is expected that nanoparticles (NPs) and molecules storing the information take more and more functions of catalysts, nanotools, and nanodevices, on the one hand, whereas hardware will become more and more changeable and take tasks of incorporation of changeable information, on the other hand.

The borderlines between hardware and software will be blurred, as well as the borderline between materials and devices. The handling of molecules for information storage is a process of chemical synthesis, at least a kind of isomerization. It is always marked by a local conversion of material, a chemical reaction. Thus, the writing of information into particle-incorporated structures is a material conversion process, at least. Stored information causes the formation or conversion of other molecules and the generation of physical signals during the readout processes. This corresponds completely to a lot of recent examples of chemical and biochemical assays and sensor procedures in which the sensing elements are also convertible. Immunoassays or DNA hybridization assays are typical examples of such more or less nonreversible recognition processes. Probably, the traditional thinking of a sensor as a complete reversible system that is generating information by a short interaction, but without changing itself, does not match with the future particle-based communication system. Any step—from the transfer of information into molecular or NP structures over the packaging into particles, the release, and transport of particles, as well as their recognition and read out by the receiver—will be a nonreversible process. Communication beads are not only passive carriers, but they undergo changes and causes changes in the receiving system as a catalyst is inducing changes in substrate substances and systems.

8.4.3   The Requirement of Convergence

The further development of information-storing particles and particle-based communication systems will be accompanied by new approaches from the storing materials, particle architectures, and reading and writing systems, on the one hand. On the other hand, it will be marked by an increasing need of standardization, conventions,. and simplifications.

It is to be assumed that certain successful basic solutions for the choice of particle materials and information-carrying molecules or nanostructures will become standards in the further evolution of particle-based communication systems. This establishing of basic standards could proceed similar to the introduction and further continuous use of the basic commands of machine codes in the

computer world and the DOS command sets as the fundament of all
following developments of computer programs.

Complexity can only be created if well-working conventions for the
basic levels are available and applied universally. The development
of communication particles has to respect this fundamental rule. It
is dictated from an economic point of view because a high diversity
in the base units would create immense costs at the more complex
levels. An efficient production and use of large sets of hierarchically
constructed communication particles are unimaginable if many
different basic principles and materials or molecule types are
applied. And it is dictated from the efficiency of development, which
means from a more evolutionary point of view. At a certain point of
development, the competition between the basic principles has to be
decided and all following steps of development have to be based on
the principles of the winners.

This convergence in fundamental aspects of technical
developments meets a complementary situation in the establishing
of fundamental basic principles in the development of life. It is not
only recommendable to follow some very important experiences
of biological evolution, but it is essential to respect them if
sustainable development should be achieved. This insight is valid
for the ecological as well as for the functional aspects of technical
developments.

The use of the principle of distributed information portions in
the form of particles is an obvious convergence to the principle of
producing of polls and seeds by plants or of offspring, in general.
In many cases, the production and distribution of large numbers
of information bundles is the easiest and most efficient way for the
distribution of this information, obviously.

In a similar way, the double strategy using large and small particles
for communication is a direct convergence with the development of
the **r** and the **k** strategy of producing offspring in living nature. Both
principles are proofed since many millions of years and confirmed
as powerful strategies under different evolutionary conditions. And
they look reasonable from a technical point of view, too.

But besides the functional aspects of convergence, ecological
sustainability is a second and still more important reason for
convergence between natural and technical communication systems.
At the moment, we are learning how we have to correct the industrial

basis for energy production in order to achieve sustainability. The so-called alternative energy sources—wind, water, waves, and photovoltaics—are exploiting the incoming energy of the sun at short timescales as living nature is doing by photosynthesis.

Behind the conversion of industrial energy production, we have to be aware that more efforts are required for putting our production of materials on the basis of renewables, as soon as possible, too. Thus, also the communication materials and systems should be based on renewable and biological compatible materials. Not only size and distribution but also the material of communication particles have to be gained from natural sources and be adapted to the environment. Particle-based communication systems will only become sustainable if they become more and more similar to micro- and nanoscaled natural objects.

Bibliography

Achatz, D.E., et al. (2009). *ChemBiochem*, **14**, 2316–2320.

Ahmad, R., et al. (2015). *Chem. Mater.*, **27**, 5464–5478.

Ahmed, S.R., et al. (2013). *J. Nanobiotechnol.*, **22**, 28.

Albers, A.E., et al. (2012). *J. Am. Chem. Soc.*, **134**, 9565–9568.

Alpsoy, L., et al. (2017). *J. Supercond. Novel Magn.*, **30**, 2797–2804.

An, K. and Hyeon, T. (2009). *Nano Today*, **4**, 359–373.

Ang, C.Y., et al. (2014). *Sci. Rep.*, **4**, 7057.

Ariga, K., et al. (2012). *NPG Asia Mater.*, **4**, doi: 10.1038/am.2012.30.

Baffou, G. and Quidant, R. (2013). *Laser Photonics Rev.*, **7**, 171–187.

Bai, M., et al. (2015). *Anal. Chem.*, **87**, 2383–2388.

Bajaj, A., et al. (2009). *Proc. Natl. Acad. Sci. U. S. A.*, **106**, 10912–10916.

Bagwe, R.P., et al. (2004). *Langmuir*, **20**, 8336–8342.

Bhunia, S.K., et al. (2014). *ACS Appl. Mater. Interfaces*, **6**, 7672–7679.

Boenemann, K., et al. (2012). *Nano-Biotechnol. Biomed. Diagnostic Res.*, **733**, 63–74.

Borges, M., et al. (2015). *RSC Adv.*, **5**, 86779.

Bouquey, M., et al. (2008). *Chem. Eng. J.*, **135**, S93–S98.

Buffi, N., et al. (2011). *Lab Chip*, **11**, 2369–2377.

Bunz, U.H.F. and Rotello, M. (2010). *Angew. Chem. Int. Ed.*, **122**, 2–15.

Burns, A., et al. (2006). *Small*, **2**, 723–726.

Cao, J., et al. (2015). *Microchim. Acta*, **182**, 385–394.

Cao, X.H., et al. (2017). *Microchim. Acta*, **184**, 3273–3279.

Carrasco, E., et al. (2015). *Adv. Funct. Mater.*, **25**, 615–626.

Cen, Y., et al. (2014). *Anal. Chem.*, **86**, 7119–7127.

Chang, Z.Q., et al. (2009). *Lab Chip,* **9** 3007–3011.

Chebl, M., et al. (2017). *Talanta*, **169**, 104–114.

Chemburu, S., et al. (2008). *J. Phys. Chem. C*, 14492–14499.

Chen, B., et al. (2013). *Nanoscale*, **5**, 8541–8549.

Chen, B., et al. (2016). *Small*, **12**, 782–792.

Chen, C.-C. V., et al. (2013). *PLoS One*, 0056125.

Chen, H.B., et al. (2017). *Biomaterials*, **144**, 42–52.

Chen, M., et al. (2014). *Adv. Mater.*, **26**, 8210–8216.

Cheng, S.-H., et al. (2010). *J. Mater. Chem.*, **20**, 6149–6157.

Chen, W.-H., et al. (2014). *Nanoscale*, **6**, 9531–9535.

Chen, Y., et al. (2013). *J. Gastroenterol.*, **48**, 809–821.

Chi, B.-Z., et al. (2017). *Biosens. Bioelectron.*, **87**, 216–221.

Chu, B., et al. (2016). *Anal. Chem.*, **88**, 9235–9242.

Csáki, A., et al. (2001). *Nucleic Acids Res.*, **29**, e81.

Cunningham, C.H., et al. (2005). *Magn. Reson. Med.*, **53**, 999–1005.

Debagge, P. and Jaschke, W. (2008). *Histochem. Cell Biol.*, **130**, 845–875.

Decher, G. (1997). *Science*, **277**, 1232–1237.

Decher, G., et al. (1992). *Thin Solid Films*, **210**, 831–835.

DelMercato, L.L., et al. (2014). *Adv. Colloid Interface Sci.*, **207**, 139–154.

Demuth, C., et al. (2016). *Appl. Microbiol. Biotechnol.*, **100**, 3853–3863.

Deng, M.L. and Wang, L.Y. (2014). *Nano Res.*, **7**, 782–793.

Diez, I. and Ras, R.H.A. (2011). *Advanced Fluorescence Reporters in Chemistry and Biology II*, Vol. 9, Springer-Verlag Berlin Heidelberg, pp. 307–332.

Discher, D.E. and Ahmed, F. (2006). *Annu. Rev. Biomed. Eng.*, **8**, 323.

Dimitriev, R.I. and Papkovsky, D.B. (2012). *Cell. Mol. Life Sci.*, **69**, 2025–2039.

Dimitriev, R.I., et al. (2015a). *Cell. Mol. Life Sci.*, **72**, 367–381.

Dimitriev, R.I., et al. (2015b). *ACS Nano*, **9**, 5275–5288.

Du, L.H., et al. (2017). *ACS Nano*, **11**, 8930–8943.

Du, P., et al. (2017). *Sens. Actuators B*, **252**, 584–591.

Duan, N., et al. (2015). *Microchim. Acta*, **182**, 917–923.

Ehgartner, J., et al. (2016). *Anal. Chem.*, **88**, 9796–9804.

Emrani, A.S., et al. (2016). *Biosens. Bioelectron.*, **79**, 288–293.

Fan, X., et al. (2016). *Chem. Commun.*, **52**, 1178–1181.

Fan, Z., et al. (2016). *Nat. Nanotechnol.*, **11**, 388.

Floris, A., et al. (2011). *Soft Matter*, **7**, 6239–6247.

Fortunato, M.E., et al. (2010). *Chem. Mater.*, **22**, 1610–1612.

Frasconi, M., et al. (2010). *Anal. Chem.*, **82**, 2512–2519.

Frenkel, J. and Doefman, J. (1930). *Nature*, **126**, 274–275.

Feng, L., et al. (2014). *Adv. Mater.*, **26**, 3926–3930.

Feng, X., et al. (2010). *ACS Appl. Mater. Interfaces*, **2**, 2429–2435.

Feng, X., et al. (2012). *Adv. Mater.*, **24**, 637–641.

Frisk, M.L., et al. (2008). *Lab Chip*, **8**, 1793–1800.

Fritzsche, W. (2001). *Rev. Mol. Biotechnol.*, **82**, 37–46.

Fritzsche, W. and Taton, T.A. (2003). *Nanotechnology*, **14**, R63–R73.

Funfak, A., et al. (2009). *Microchim. Acta*, **164**, 279–286.

Gao, X., et al. (2002). *J. Biomed. Opt.*, **7**, 532–537.

Glynne-Jones, P., et al. (2010). *Ultrasonics*, **50**, 235–239.

Gogoi, M. and Deb, P. (2014). *Appl. Surf. Sci.*, **198**, 130–136.

Gontero, D., et al. (2017). *Microchem. J.*, **130**, 316–328.

Guice, K.B., et al. (2005). *J. Biomed. Opt.*, **10**, 064031.

Gupta, A.K. and Gupta, M. (2005). *Biomaterials*, **26**, 3995–4021.

Hanel, C. and Gauglitz, G. (2002). *Anal. Bioanal. Chem.*, **372**, 91–100.

He, C.S., et al. (2010). *J. Mater. Chem.*, **20**, 10755–10764.

He, H., et al. (2013). *Anal. Chem.*, **85**, 4546–4553.

Hoffmann, K., et al. (2013). *ACS Nano*, **7**, 6674–6684.

Horka, M., et al. (2016). *Anal. Chem.*, **88**, 12006–12012.

Huang, P., et al. (2015). *Anal. Chem.*, **87**, 6834–6841.

Jiang, J., et al. (2012). *Chem. Commun.*, **48**, 9634–9636.

Jiang, Y.Y. and Pu, K.Y. (2017). *Small*, **13**, UNSP 1700710.

Jin, B.R., et al. (2017). *Biosens. Bioelectron.*, **90**, 525–533.

Jokerst, J.V., et al. (2012). *ACS Nano*, **6**, 10366–10377.

Khodakovskaya, M.V., et al. (2011). *Proc. Natl. Acad. Sci. U. S. A.*, **108**, 1028–1033.

Kim, J.W., et al. (2009). *Nat. Nanotechnol.*, **4**, 688–694.

Kneipp, J. (2017). *ACS Nano*, **11**, 1136–1141.

Kneipp, K., et al. (2015). *Chem. Sci.*, **6**, 2721–2726.

Köhler, J.M., et al. (2013). *Anal. Chem.*, **85**, 313–318.

Konno, T., et al. (2006). *J. Biomater. Sci. Polym. Ed.*, **17**, 1347–1357.

Koo, Y.E.L., et al. (2004). *Anal. Chem.*, **76**, 2498–2505.

Korzeniowska, B., et al. (2013). *Nanotechnology*, **24**, 442002.

Kreft, O., et al. (2002). *J. Mater. Chem.*, **17**, 4471–4476.

Kukula, H., et al. (2002). *J. Am. Chem. Soc.*, **124**, 1658.

Kulkarni, B. and Jayakannan, M. (2017). *ACS Biomater.*, **3**, 2185–2197.

Lam, A.T., et al. (2017). *Lab Chip*, **17**, 1442–1451.

Lay, A., et al. (2017). *Nano Lett.*, **17**, 4172–4177.

Lee, I., et al. (2014). *Appl. Mater. Interfaces*, **6**, 17151–17156.

Lee, J.-H., et al. (2009). *Nanotechnology*, **20**, 355102.

Lensen, D., et al. (2008). *Macromol. Biosci.*, **8**, 991–1005.

Lian, W., et al. (2004). *Anal. Biochem.*, **334**, 135–144.

Liang, K., et al. (2014). *Adv. Healthcare Mater.*, **3**, 1551–1554.

Li, H., et al. (2010). *Angew. Chem. Int. Ed.*, **49**, 4430–4434.

Li, H., et al. (2015). *J. Mater. Chem. B*, **3**, 1193–1197.

Li., K. and Liu, B. (2014). *Chem. Soc. Rev.*, **43**, 6570–6597.

Li, N., et al. (2014). *Chem. Eur. J.*, **20**, 16488–16491.

Li, X., et al. (2017). *Chem. Eng. J.*, **326**, 1058–1065.

Li, Y., et al. (2010). *Gold Bull.*, **43**, 29–40.

Li, Y., et al. (2017). *J. Controlled Release*, **258**, 95–107.

Lin, C.-A. J., et al. (2009). *J. Med. Biol. Eng.*, **29**, 276–283.

Lin, F.E., et al. (2016). *Talanta*, **158**, 292–298.

Liu, H., et al. (2013). *Nanoscale*, **5**, 9340–9347.

Liu, J., et al. (2013). *Polym. Chem.*, **4**, 4326–4334.

Liu, P., et al. (2015). *ACS Appl. Mater. Interfaces*, **7**, 6754–6763.

Liu, R., et al. (2009). *Angew. Chem. Int. Ed.*, **48**, 4598–4601.

Liu, R., et al. (2017). *Appl. Mater. Interfaces*, **9**, 3006–3015.

Lu, J.Z. and Rosenzweig, Z. (2000). *Fresenius J. Anal. Chem.*, **366**, 569–575.

Lupitskyy, R., Motornov, M. and Minko, S. (2008). *Langmuir*, **24**, 8976–8980.

Ma, X.H., et al. (2015). *Colloids Surf. B*, **126**, 44–49.

Mahajan, P.G., et al. (2016). *J. Fluoresc.*, **26**, 1467–1478.

Mahtab, F., et al. (2011). *Small*, **7**, 1448–1455.

Makwana, B.A., et al. (2017). *Sens. Actuators B*, **246**, 686–695.

März, A., et al. (2011). *Lab Chip*, **11**, 3584–3592.

März, A., et al. (2012). *Anal. Bioanal. Chem.*, **402**, 2625–2631.

Meldal, M. and Christensen, S.F. (2010). *Angew. Chem. Int. Ed.*, **122**, 1–5.

Motornov, M., et al. (2010). *Prog. Polym. Sci.*, **35**, 174–211.

Mühlig, A., et al. (2016). *Anal. Chem.*, **88**, 7998–8004.

Muthiah, M., et al. (2013). *Carbohydr. Polym.*, **92**, 1586–1595.

Nagl, S. and Wolfbeis, O.S. (2007). *Analyst*, **132**, 507–511.

Nie, Q., et al. (2006). *J. Mater. Chem.*, **16**, 546–549.

Ow, H., et al. (2005). *Nano Lett.*, **5**, 113–117.

Pahlow, S., et al. (2015). *Adv. Drug Deliv. Rev.*, **89**, 105–120.

Patel, D., et al. (2010). *Biomaterials*, **31**, 2866–2873.

Patil, K.S., et al. (2017). *Spectrochim. Acta A*, **170**, 131–137.

Peng, J., et al. (2007). *Talanta*, **71**, 833–840.

Peper, S., et al. (2003). *Anal. Chim. Acta*, **500**, 127–136.

Petrizza, L., et al. (2016). *RSC Adv.*, **6**, 104164–104172.

Petry, R., et al. (2003). *ChemPhysChem*, **4**, 14–30.

Polyak, B. and Friedman, G. (2009). *Expert Opin. Drug Deliv.*, **6**, 53–70.

Potara, M., et al. (2013). *Nanoscale*, **5**, 6013–6022.

Qian, C.-G., et al. (2015). *Appl. Mater. Interfaces*, **7**, 18581–18589.

Rad, J.K., et al. (2016). *Polymer*, **98**, 263–269.

Rahman, M., et al. (2015). *Curr. Org. Chem.*, **19**, 982–990.

Resch-Genger, U., et al. (2008). *Nat. Methods*, **5**, 763–775.

Riskin, M., et al. (2010). *Chem. Eur. J.*, **16**, 7114–7120.

Rivera-Gil, P., et al. (2013). *Accounts Chem. Res.*, **46**, 743–749.

Rocha, U., et al. (2016). *J. Lumin.*, **175**, 149–157.

Sakulkhu, U., et al. (2015). *Biomater. Sci.*, **3**, 265–278.

Samal, A.K., et al. (2013). *Langmuir*, **29**, 15076–15082.

Sarkar, K., et al. (2009). *Adv. Funct. Mater.*, **19**, 223–234.

Scarmagnani, S., et al. (2008). *J. Mater. Chem.*, **18**, 5063–5071.

Scheucher, E., et al. (2015). *Microsyst. Nanoeng.*, **1**, UNSP 15026.

Schulz, A. and McDonagh, C. (2012). *Soft Matter*, **8**, 2579–2585.

Seo, S., et al. (2010). *Eur. J. Inorg. Chem.*, 843–847.

Serra, C., et al. (2007). *Langmuir*, **23**, 7745–7750.

Serra, C.A. and Chang, Z.Q. (2008). *Chem. Eng. Technol.*, **31**, 1099–1115.

Serra, C.A., et al. (2013a). *J. Flow Chem.*, **3**, 66–75.

Serra, C.A. et al. (2013b). *Macromol. Reaction Eng.,* **7**, 414–439.

Shaskov, E.V., et al. (2008). *Nano Lett.*, **8**, 3953–3958.

Shi, B., et al. (2016). *Biosens. Bioelectron.*, **82**, 233–239.

Shi, J., et al. (2017). *Biosens. Bioelectron.*, **93**, 182–188.

Shi, J.Y., et al. (2015). *J. Mater. Chem. B*, **3**, 6989–7005.

Shirivand, G., et al. (2017). *Sens. Actuators B*, 244–252.

Singh, V. and Mishra, A.K. (2016). *Sens. Actuators B*, **227**, 467–474.

Smith, J.E., et al. (2007). *Anal. Chem.*,**79**, 3075–3082.

Solberg, N.O., et al. (2014). *J. Alzheimers Dis.*, **40**, 191–212.

Song, Y.N., et al. (2014). *Biomaterials*, **35**, 2961–2970.

Stanca, S.E., et al. (2010). *Nanotechnology*, **21**, 055501.

Steinbrück, A., et al. (2011). *Anal. Bioanal. Chem.*, **401**, 1241–1249.

Strehle, K.R., et al. (2007). *Anal. Chem.*, **79**, 1542–1547.

Stromer, B.S. and Kumar, C.V. (2017). *Adv. Funct. Mater.*, **27**, 1603874.

Suda, Y., et al. (2002). *Thin Solid Films*, **415**, 15–20.

Su-dai, M., et al. (2016). *Anal. Chem.*, **88**, 10474–10481.

Sukhorukov, G.B., et al. (2007). *Small*, **3**, 944–955.

Sun, C.L., et al. (2015). *Anal. Bioanal. Chem.*, **407**, 8883–8892.

Sun, S., et al. (2015). *Methods Appl. Fluoresc.*, **3**, 034002.

Sung, T.W. and Lo, Y.L. (2012). *Sens. Actuators B*, **165**, 119–125.

Thatai, S., et al. (2014). *Microchem. J.*, **113**, 77–82.

Thomas, R.G., et al. (2017). *Macromol. Res.*, **25**, 446–451.

Tong, L., et al. (2009). *Photochem. Photobiol.*, **85**, 21–32.

Toole, B.P. (2004). *Nat. Rev. Cancer*, **4**, 528.

Tsagkatakis, I., et al. (2001). *Anal. Chem.*, **73**, 6083–6087.

Tu, Z.Z., et al. (2017). *J. Nanosci. Nanotechnol.*, **17**, 123–132.

Van Blaaderen, A. and Vrij, A. (1992). *Langmuir*, **8**, 2921–2931.

Verma, S.K., et al. (2016). *Adv. Funct. Mater.*, **33**, 6015–6024.

Visaveliya, N., et al. (2015a). *ACS Appl. Mater. Interfaces*, **7**, 10742–10754.

Visaveliya, N., et al. (2015b). *Chem. Eng. Technol.*, **38**, 1144–1149.

Visaveliya, N. and Köhler, J.M. (2014). *ACS Appl. Mater. Interfaces*, **6**, 11254–11264.

Visaveliya, N. and Köhler, J.M. (2015). *J. Mater. Chem. C*, **3**, 844–853.

Visaveliya, N.R., et al. (2017). *Macromol. Chem. Phys.*, **218**, 1700261.

Von der Lühe, M., et al. (2015). *RSC Adv.*, **5**, 31920–31929.

Wackerlig, J. and Lieberzeit, P.A. (2015). *Sens. Actuators B*, **207**, 144–157.

Walter, A., et al. (2011). *Lab Chip*, **11**, 1013–1021.

Wang, F. and Liu, X.G. (2008). *J. Am. Chem. Soc.*, **130**, 5642–5643.

Wang, F., et al. (2010). *Nature*, **463**, 1061–1065.

Wang, H., et al. (2017). *Polym. Chem.*, **8**, 5795–5802.

Wang, H.-B., et al. (2015). *Sens. Actuators B*, **220**, 146–153.

Wang, J., et al. (2011). *J. Mater. Chem.*, **21**, 18696–18703.

Wang, K., et al. (2013). *Acc. Chem. Res.*, **7**, 1367–1376.

Wang, L. and Tan, W. (2006). *Nano Lett.*, **6**, 84–88.

Wang, L. and Li, Y. (2006). *Nano Lett.*, **6**, 1645–1649.

Wang, W., et al. (2014). *Acc. Chem. Res.*, **47**, 373–384.

Wang, X., et al. (2007). *Bioconjugate Chem.*, **18**, 297–301.

Wang, X., et al. (2011). *Chem. Commun.*, **47**, 4261–4263.

Wang, Z., et al. (2010). *Microbiol. Methods*, **83**, 179–184.

Wang, Z.H., et al. (2013). *Org. Biomol. Chem.*, **11**, 3159–3167.

Weber, P., et al. (2018). *Sens. Actuators B*, **267**, 26–33.

Widder, K.J., et al. (1980). *Cancer Res.*, **40**, 3512–3517.

Willner, I. and Willner, B. (2002). *Pure Appl. Chem.*, **74**, 1773–1783.

Wu, J.Y. and Hsu, K.Y. (2015). *J. Mol. Struct.*, **1099**, 142–148.

Wu, W.T., et al. (2012). *Biomaterials*, **33**, 7115–7125.

Xiong, B., et al. (2017). *ACS Nano*, **11**, 541–548.

Xu, J., et al. (2016). *Biosens. Bioelectron.*, **75**, 1–7.

Xu, J.W., et al. (2001). *Anal. Chem.*, **73**, 4124–4133.

Xu, L.Q., et al. (2012). *Polym. Chem.*, **3**, 2444–2450.

Xu, S., et al. (2016). *Anal. Chem.*, **88**, 7853–7857.

Yang, K., et al. (2014). *Theranostics*, **4**, 134–141.

Yang, S.W. and Vosch, T. (2011). *Anal. Chem.*, **83**, 6935–6939.

Yang, X.M., et al. (2009). *Wiley Interdiscip. Rev. Nanomed. Nanobiotechnol.*, **1**, 360–368.

Yang, Z.C., et al. (2011). *Chem. Comm.*, **47**, 11615–11617.

Ye, N., et al. (2007). *Anal. Chim. Acta*, **596**, 195–200.

Yin, S., et al. (2005). *Biophys. J.*, **88**, 1489–1495.

Yow, H.N. and Routh, A.F. (2006). *Soft Matter*, **2**, 940.

Yu, C., et al. (2013). *Chem. Commun.*, **49**, 403–405.

Yu, X., et al. (2014). *RSC Adv.*, **4**, 23571–23579.

Yuan, Y., et al. (2015). *Anal. Chem.*, **87**, 6180–6185.

Zamaleeva, A.I., et al. (2015). *Sensors*, **15**, 24662–24680.

Zan, X., et al. (2015). *Langmuir*, **31**, 7601–7608.

Zhang, B., et al. (2012). *Appl. Mater. Interfaces*, **4**, 6479–6486.

Zhang, K., et al. (2016). *Mater. Lett.*, **172**, 112–115.

Zhang, K.Y., et al. (2015). *Chem. Sci.*, **6**, 301–307.

Zhang, P., et al. (2014). *Macromol. Biosci.*, **14**, 1495–1504.

Zhao, S., et al. (2017). *J. Mater. Chem. B*, **5**, 3651–3657.

Zhao, T. B., et al. (2017). *Nanoscale*, **9**, 9841–9847.

Zhou, B., et al. (2000). *J. Biol. Chem.*, **275**, 37733.

Zhou, Y.W., et al. (2017). *Nanoscale*, **9**, 12746–12754.

Zhu, K.N., et al. (2017). *Macromolecules*, **50**, 1113–1125.

Zhu, L., et al. (2005). *J. Am. Chem. Soc.*, **127**, 8968–8970.

Zhu, M.-Q., et al. (2005). *J. Am. Chem. Soc.*, **127**, 8968–8970.

Zhu, M.-Q., et al. (2006). *JACS*, **128**, 4303–4309.

Zhu, M.-Q., et al. (2011). *JACS*, **133**, 365–372.

Index